Explorations in the Icy North

SCIENCE AND CULTURE IN THE
NINETEENTH CENTURY

Bernard Lightman, Editor

EXPLORATIONS *in the* ICY NORTH

HOW
TRAVEL NARRATIVES
SHAPED ARCTIC SCIENCE
IN THE NINETEENTH CENTURY

Nanna Katrine Lüders Kaalund

UNIVERSITY *of* PITTSBURGH PRESS

Published by the University of Pittsburgh Press, Pittsburgh, Pa., 15260

Copyright © 2021, University of Pittsburgh Press

All rights reserved

Manufactured in the United States of America

Printed on acid-free paper

10 9 8 7 6 5 4 3 2 1

Cataloging-in-Publication data is available from the Library of Congress

ISBN 13: 978-0-8229-4659-5

ISBN 10: 0-8229-4659-9

Cover art: Cover image to sheet music for G. Jervis Rubini's *The Artic Waltzes*, 1876.

Cover design: Alex Wolfe

This is for my son Magnus, who makes every day better.

CONTENTS

ACKNOWLEDGMENTS

The journey from initial idea to published book would not have been possible without the help, support, and advice of several people. First and foremost, I am indebted to Bernie Lightman for his continuing guidance and encouragement, both academically and personally. I could not have wished for a better mentor than Bernie. I would also like to thank wholeheartedly the wider academic community at York University, Toronto. The PhD program in science and technology studies at York University was a wonderful place for developing my research and building friendships. I am very thankful to Colin Coates and Ernst Hamm, and to Deborah Neill, Katharine Anderson, Aryn Martin, Alexandra Rutherford, James Elwick, and Adrian Shubert. I am also indebted to Janet Browne for fantastic feedback and support. The importance of their guidance during the process of writing this book cannot be overstated. I received funding from the Ontario Graduate Scholarship scheme to support my doctoral studies, for which I am very grateful.

My second academic home is in Leeds, where I first became interested in nineteenth-century history of science during my MA. The Centre for History and Philosophy of Science made the (perhaps questionable) decision of hiring me as a lecturer in the history of science communication after I completed my PhD, and I loved every second of working there. I am particularly thankful to Jonathan Topham and Graeme Gooday, who, although they have known me since I was an MA student, have continued to be wonderfully supportive friends and mentors. I am also thankful to the wider research community at Leeds, including Jamie Stark, Adrian Wilson, Greg Radick, Becky Bowd, Thomas Brouwer, Liz Bruton, and Helen Steward.

I researched and wrote much of this book at the Scott Polar Research Institute in Cambridge. I am very thankful to Michael Bravo and Richard Powell, for their support and advice at different points during the journey of completing this book. The Scott Polar Research Institute is a wonderful place to work, and I am fortunate to have had access to the Polar Library and the Thomas H. Manning Polar Archives. My eternal gratitude extends to Naomi Boneham, Laura Ibbett,

Frances Marsh, and Peter Lund for their help with accessing the library and archive, and for their insights into polar history.

My thanks also extend to the University of Pittsburgh Press, in particular to Abby Collier. Abby is a truly incredible editor and has made the process of finishing this book very enjoyable. I am also indebted to the insightful comments of the two anonymous reviewers. Their suggestions strengthened the book immensely, and I hope they are as pleased with the final result as I am.

Finally, I am thankful to my friends and family. Throughout the years, many people have helped and supported me, and made it possible for me to both start and finish this project. I am lucky to have the continued friendships of Sara Kvist, Geoff Belknap, Helen Coskeran, Jon Livingstone-Banks, Michaela Livingstone-Banks, Sidsel Wittendorff Sørensen, and Christie Busich. I am very grateful to my family, in particular my parents and siblings, for believing in me and supporting me in my travels from Denmark to England, before going to Canada, and then back to England again. Finally, I would like to thank my husband, Efram Sera-Shriar, without whom none of this would have been possible.

Explorations in the Icy North

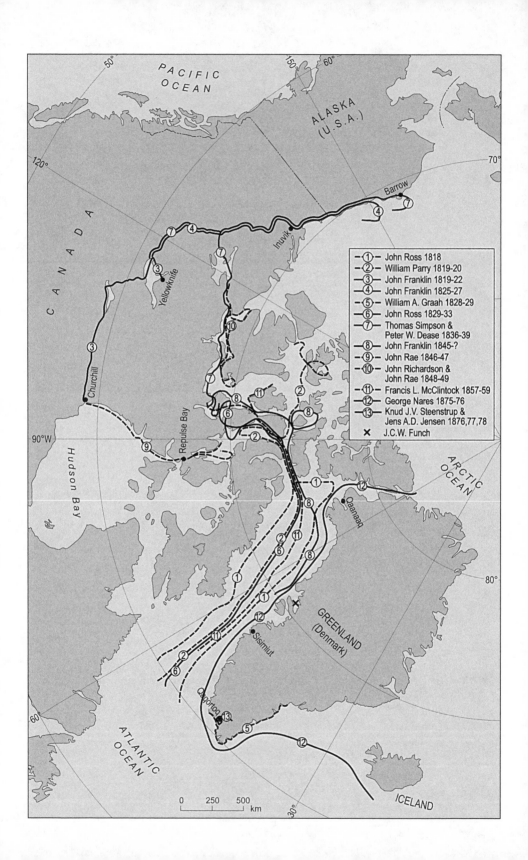

PACIFIC OCEAN

ALASKA (U.S.A.)

Barrow

Inuvik

CANADA

Yellowknife

Churchill

Hudson Bay

Repulse Bay

ATLANTIC OCEAN

ARCTIC OCEAN

Qaanaaq

GREENLAND (Denmark)

Sisimiut

Qaqortoq

ICELAND

—①— John Ross 1818
—②— William Parry 1819-20
—③— John Franklin 1819-22
—④— John Franklin 1825-27
—⑤— William A. Graah 1828-29
—⑥— John Ross 1829-33
—⑦— Thomas Simpson & Peter W. Dease 1836-39
—⑧— John Franklin 1845-?
—⑨— John Rae 1846-47
—⑩— John Richardson & John Rae 1848-49
—⑪— Francis L. McClintock 1857-59
—⑫— George Nares 1875-76
—⑬— Knud J.V. Steenstrup & Jens A.D. Jensen 1876,77,78
✕   J.C.W. Funch

0   250   500 km

INTRODUCTION

By the copious details they [expedition narratives] embrace, in every branch of astronomical and nautical science, of geography, meteorology, and other physical researchers,—the charts and prints by which they are illustrated—they are made highly valuable to the man of science and taste, and well adapted for public libraries.

— John Barrow, *Voyages of Discovery and Research within the Arctic Regions*, 1846

What was the purpose of Arctic exploration, and why publish accounts of the voyages? These were the critical questions the second secretary to the British Admiralty John Barrow (1764–1848) sought to address in his book *Voyages of Discovery and Research within the Arctic Regions* (1846). Barrow had been a key advocate of British government–funded Arctic exploratory voyages over the previous four decades, and though none had succeeded in finding the fabled Northwest Passage, his book was a passionate argument for the importance of Arctic exploration. The many past Arctic expeditions may not have found a trading route to the Pacific, but landscapes, Barrow argued, "must be traced" as geographical surveying cannot be conjectured. Further to geographical discovery, Barrow wrote, the "acquisition of knowledge is the groundwork" for the instructions given to the explorers, directing them to undertake "constant observations" for the "advancement of every branch of science."[1] These results were published in scientific journals, but another genre arguably reached a broader audience: travel literature.

◄ FIGURE I.1. Map showing the approximate routes taken by the main expeditions discussed in the following four chapters. The map was kindly produced for this book by Philip Stickler, Department of Geography, University of Cambridge.

In the nineteenth century travel literature served multiple purposes and therefore had multiple audiences. It was an important evidentiary resource for many scientific disciplines, and it was one of the primary sources through which diverse groups of readers could learn about parts of the world they would never visit themselves. It was a popular genre, one that spoke of distant lands, strange animals and plants, and unfamiliar, exotic cultures. They were intended to be captivating accounts, typically of heroism in the face of danger in unknown regions, and the diary format used in most narratives invited the reader to join in the discoveries. While the decision was often taken to place the majority of the scientific measurements and observations in appendixes to the narratives, the day-to-day format of the main body of the text included descriptions of scientific investigations as well as the experiences of the explorers. In fact, Barrow argued that this feature of Arctic narratives was problematic, as it contributed to the high cost and extensive length of the books.[2] It was also problematic in subtler ways. The multiple functions of the travel writing format posed unique challenges to the authors. The diary format suggested that this was an unedited and direct account of the Arctic; so when individual aspects of the geographical, scientific, and experiential parts of a narrative were questioned, it had the potential to delegitimize not only the results, but the explorer as well. It was never a given that narratives were accepted as a true account of the Arctic. Their veracity was linked to the author, as well as to the surrounding circumstances of the expedition and the textual strategies employed within them. This book concerns this process and asks questions about how explorers constructed the Arctic, their scientific practices, and themselves in their travel narratives.

Arctic explorers were expected to undertake investigations into a wide range of scientific areas, including geology, anthropology, ethnography, medicine, geography, hydrography, meteorology, magnetic and astronomical science, botany, natural history, and glaciology. Geography was, and is, a branch of science, but geographical surveying was treated and evaluated separately from results relating to other scientific fields. In sum, explorers were expected to function as jacks of all trades. The Arctic was a particularly intriguing site for many scientific disciplines, as explorers would encounter extreme weather, rugged terrain, and unusual fauna and flora. It was hoped that by studying the Arctic, it would be possible to elucidate not only the resources available in the region, but also add to the general understanding of the climate. The definition of *climate* as associated primarily with the atmospheric sciences is relatively new; it used to be related to a much broader set of issues, including health, geography, economy, and racial concerns—all of which are reflected in the travel narratives.[3] This was not limited to the Arctic, but was reflected in imperial practices in other areas as well. Significantly, the historian of science Katharine Anderson has shown how British imperialists in India perceived the region as a "natural laboratory for meteorology" because it "seemed to hold the key to unravelling the laws of the atmosphere."[4] In the same way, the Arctic was also treated as a laboratory. Scientific experiments were made locally, with the intent of applying the results globally.

Take for example the Scottish physicist and proponent of Arctic explorations, Balfour Stewart (1828–1887). According to him, Arctic research was essential because it was an important example of what he referred to as "cosmical science." For Stewart, cosmical science (what we might call geophysics today) referred to studies of the relation between solar disturbances and meteorological changes. Past breakthroughs in astronomy, such as that of Johannes Kepler and Isaac Newton, Stewart argued, were due to the type of "laborious and long-continued observations" which could be organized only by the government. Arctic whalers and merchants could not be relied upon to undertake such observations. Stewart's line of reasoning combined the ethos of an all-encompassing study of the Earth with a hierarchical view of who could provide observations for these studies. He noted that "we have before us the splendid possibility of predicting the nature of seasons; but surely we cannot expect that nature, who is usually so reticent, will disclose her secrets to a nation or a race who will not take reasonable trouble to complete their knowledge of the physics of the earth?"[5] For people like Stewart, the results from Arctic explorations were worth their effort exactly because it was difficult.

Such lofty goals and mission statements were one thing, but practice was something quite different. There was rarely a correlation between what it was hoped explorers could achieve geographically and scientifically and what they actually produced. To what extent it was possible to control the unpredictability of what explorers encountered in the Arctic was a key problem. The scientific research of Arctic explorers and the type of scientific knowledge that was produced depended on the abilities and interests of the crew, as well as the luck of the expeditions. Perhaps counterintuitively, misfortune in geographical surveying could mean a boost in scientific results: for example, having your ship caught in ice for an extended period gave you a lot of free time to undertake scientific investigations and allowed explorers to build magnetic observatories or carry out measurements of the behavior and movements of ice. As a field site the Arctic was not easily controlled, and attempts to mitigate this uncertainty reveal broader concerns about the production of field science in the nineteenth century. While the century is often described as a period of disciplinary formation, science in the Arctic followed a distinctive path. The knowledge produced there added to a broad range of scientific fields, rather than developing a distinct Arctic science. This was the case until the First International Polar Year (IPY) between 1882 and 1883, when countries came together in a concerted effort to establish a unified method for scientific research in both the Arctic and Antarctica. That is, a transition in the established scientific practices took place, from a focus on general scientific investigations in the Arctic to a more coherent Arctic science. As the Royal Navy lieutenant George T. Temple argued at the Annual Meeting for the British Association for the Advancement of Science in 1882, the IPY "marked a fresh point of departure in Polar investigation, which might now be considered as an accepted branch of study."[6]

When scientific practices in the Arctic became more formalized during the First IPY, the associations of who were authoritative observers of Arctic phenomena changed. As an anonymous commentator wrote in the London newspaper the *Standard* in 1882, the IPY was a type of Arctic work "in a shape so different from the form it has taken during the last three centuries that the 'explorers' of 1882 would be scarcely recognized by Barents, Baffin, Hudson, Frobisher, Parry, or Franklin, as members of their famous brotherhood."[7] I will explore these transitions as reflected in the travel narratives. Here I am concerned with the textual performance of this "famous brotherhood," as it was described in the *Standard*, and how the concept of the Arctic explorer shaped and was shaped by changing notions of scientific fieldwork and imperial and financial interests.

In this book, I employ broad definitions for terms such as *travel literature, narrative of exploration*, and *explorer*, and use *travel narrative, travel writing*, and *travel literature* interchangeably when referring to narratives from both large- and small-scale expeditions, as well the texts produced by more settled travelers such as missionaries. The identities of explorers and their organizing bodies shaped the expeditions, and this influenced the representation of the explorers themselves, the ventures, and the science they produced. Travel narratives also reflect the complex relationship between explorers and imperial projects. As the historian Michael Bravo has argued, "The field of postcolonial literature has taught us to attend to the narrative strategies that produce cosmopolitan authorship and authority."[8] That is, by paying close attention to the narrative strategies employed in travel accounts, especially in relation to scientific practices, we can begin to unravel the processes by which the Arctic was constructed within the European and North American colonial discourses. It is important to emphasize that the Arctic, as known to Europeans and Euro-Americans, was a cultural construct. It was continuously reimagined through different venues, not only the exploration accounts of European naval men. Expanding how we define and use the category of "explorer," allows us to see the multiple voices through which this construction of the Arctic took place.

At the same time the focus on exploration ventures and the narratives that account for them is inherently problematic, as it runs the risk of portraying the Arctic, and the Indigenous peoples there, as existing or mattering only insofar as they came into contact with foreign explorers. This, I believe, is an ingrained problem in a considerable amount of Arctic historical and contemporary research: much of it has focused on singular nations or individual European and Euro-American explorers and explorations. I acknowledge that I am also applying nation-focused and Eurocentric ways of thinking when centering my study around Danish, Canadian, and British explorations. However, I hope to partially make up for this limitation: First, I aim to untangle the processes by which the Arctic and the experiences of Arctic exploration were created and reconstructed within the confines of specific visions of what it meant to be an authoritative observer, as seen through the travel narratives. In doing so, I put aside what explorers and organizers hoped to emphasize, namely the "moment of discovery." If we take away the claims to geographical discovery and scientific achievements, it becomes clearer how Arctic exploration was never simply the work of individual figures. This deconstruction of the persona of the Arctic explorer—the white male hero—is, I believe, an important part of understanding the relationship between imperialism and field science in the nineteenth century. Second, I aim to counter the traditional Eurocentric focus by highlighting the inherent international nature of Arctic exploration.

In the following four chapters, I examine the making and communicating of knowledge about the Arctic between 1818 and the First IPY through a study of travel literature in the Danish, British, and Canadian contexts. This book is not a study of how scientific achievements in the Arctic contributed to the disciplinary formation of scientific fields in the metropole. I hope to show a different story. I am concerned with the practices of writing the Arctic experience, especially the relationship between science and the strategies for constructing a trustworthy narrative voice. I focus on the intersection of science and print to highlight the role of exploration in shaping nineteenth-century science, and reveal changes in ideas about what it meant to be an authoritative observer of natural phenomena. Such shifts were linked to tensions in imperial ambitions, national identities, and international collaborations. I combine four broad historiographical themes in order to complicate our understanding of scientific practices in the Arctic and the various sociopolitical factors that shaped that construction: the intersection of imperialism and science, the identity of the explorer, the role of travel narratives in shaping knowledge about the Arctic, and the significance of applying transnational methods to what had typically been perceived as nationalistic ventures. I will show that European and Euro-American perceptions of the Arctic, scientific practices in the Arctic, and the character of the Arctic explorer were all constructed simultaneously through the narratives and by the reception of their accounts.

IMPERIALISM, INDIGENEITY, AND SCIENCE

To tell the story of science in the Arctic, we have to engage with European and North American imperial practices and policies. The historical relationship between imperialism, science, and international collaboration is complicated and extremely violent. As with other European and North American imperial ventures throughout the world, explorers in the Arctic claimed and discovered areas that were already inhabited.[9] Although a general survey of historical accounts might give the impression that Arctic exploration was mainly the result of the zeal and bravery of a few heroic men, these so-called explorers were never alone on the ice. The success of expeditions fundamentally relied on the help of Indigenous peoples. This labor included, but was not limited to, gathering food, building shelter, and finding necessary resources such as fuels. Their assistance as guides, translators, and dogsled drivers was equally important. This part of the interactions between Indigenous peoples and the European and Euro-American explorers was usually acknowledged explicitly in travel narratives, with one major caveat: the support was presented as a type of manual labor that was nonessential to the official duties of the expeditions.

What the explorers hid from their public accounts was the fact that they drew heavily on Indigenous knowledge about the Arctic and relied on their expertise to fulfill many of the official duties of the expeditions. These duties were broad in scope and ranged from geographical surveying to the collection of natural history specimens. European and Euro-American knowledge was transformed both conceptually and empirically by Indigenous knowledge; however, this was not a simple process of information transfer between two separate, binary groups. I draw in particular on the insights of the historical geographers Felix Driver and Lowri Jones, who showed in their *Hidden Histories of Exploration* exhibition at the Royal Geographical Society (2009) that exploration knowledge at its core was the product of labor that was coproduced by Europeans and extra-Europeans. It is useful to think of this coproduction of knowledge as taking place in what the historian Mary Louise Pratt has termed the "contact zone." The contact zone, Pratt argues, is "the space in which peoples geographically and historically separated come into contact with each other and establish on-going relations, usually involving conditions of coercion, radical inequality, and intractable conflict... often within radically asymmetrical relations of power." Providing a similar analysis of encounters, the historian Stuart Schwartz has argued that an "implicit ethnography" existed within encounters during European expansion. It was ethnography, he argues, because understanding the "other" is the product of observing, reporting, and reflecting, which in turn also shape understandings of the self.[10] Reports of encounters therefore, tell us both about the observer and the observed. As I show throughout the case studies in this book, many explorers were highly attuned to their reliance on the labor of Indigenous peoples. Any power dynamic or personal relationship established in the contact zone was continually renegotiated at different points during the expeditions. This is especially pertinent when studying the Arctic, where explorations, colonialism, and scientific pursuits were characterized both by friendly collaboration, indifference, and extreme coercion and exploitation.

Just as there is no unified concept of "the local," so there is also no one singular colonial culture, discourse, or experience. This is particularly significant because cross-cultural encounters and their representations were inherently tied up with preconceived perceptions that were culturally and temporally specific. A unifying feature is that exploration was part of the process of possessing and tracing the physical landscape for imperial purposes. In addition, bringing to the fore the financial considerations involved in expeditions organized by, or in conjunction with, trading companies, helps to elucidate the differences and similarities between expeditions organized by different types of patrons. This includes those organized by trading companies and private funders, where the potential or desire for financial gain was entangled with scientific investigations in complicated ways.[11] Financial considerations were hard to overlook, as the Arctic afforded—or appeared to afford—opportunities to exploit natural resources for economic gain. Finding a Northwest Passage was also grounded, at least initially, in financial concerns, as it potentially could provide an important trading route to the Pacific.

Ownership, right to resources, and potential trading routes were main motivators in the organization of many Arctic expeditions. As the Kongelige Grønlandske Handel (KGH) worked with the Danish Crown, the Hudson's Bay Company (HBC) collaborated with the British Navy to survey the North American Arctic in overland expeditions. While Britain was not experiencing wars within its own borders, the British Empire was engaged in conflicts throughout the world, including, but not limited to, the First Opium War (1839–42) and the First (1839–1842) and Second (1848–49) Anglo-Afghan Wars. There were also conflict and political unrest in Canada. In 1837 there were rebellions in both Lower Canada (present-day Quebec) and Upper Canada (present-day Ontario). While the British government defeated the rebellions, they ultimately led to greater autonomy in the region, and in 1841 Lower and Upper Canada were combined under the United Province of Canada. In Canada science in the Arctic was a way to establish sovereignty in the region and to confirm and build a Canadian national settler identity. In particular, the historian of science Trevor Levere has emphasized the significance of national concerns, international cooperation, and national rivalries in sending out explorers to the Arctic in the British-Canadian context.[12] The HBC was an important patron of science, as scientific activity could be strategically framed as a way of bettering its troubled reputation. For example, the HBC was involved with learned societies in Canada, Britain, and the United States and used scientific field research as a way to strengthen its reputation. This countered the many critiques that questioned the validity of the HBC's trading monopoly, as well as its treatment of Indigenous peoples. We see the same with the KGH in Greenland, where the trading company supported and controlled explorers and settlers. The KGH and the HBC both enjoyed a monopoly on trade, and their efforts to maintain this control influenced the trajectory of several Arctic explorations. British North America covered a vastly larger area than the United Province of Canada, from the Atlantic to the Great Lakes, while the HBC still enjoyed a trade monopoly and control over Ruperts Land. The areas of interest in the search for a Northwest Passage, specifically north of Davis Strait and Baffin Bay, were outside the authority of the HBC. In these expeditions, the company and the Royal Navy explorers relied on the assistance of fur traders and Indigenous peoples. Similarly, the overland and littoral expeditions backed by the Danish Crown and the KGH extended into areas outside of their direct authority.[13]

In the Danish context, discovering traces of the lost Nordic tribe was a key concern at the beginning of the nineteenth century.[14] Proof of their continued existence would support the Danish claim to the area, something the newly sovereign Norway contested. In contrast to Britain's enthusiasm for the Arctic during the first half of the nineteenth century, the Danish Crown and the KGH had difficulties organizing expeditions for any purpose, be it geographical or scientific, because there was an acute lack of funds available in Denmark at that time. This financial predicament spurred interest in cataloging the potential resources in Greenland, which could be exploited for financial gain. For the Danish Crown and the KGH, as for the HBC, the links between knowledge about the Arctic and economic and imperial concerns are evident. The HBC struggled with a large debt and a new organizational structure in the period following the merger between the HBC and the North-West Company (NWC).[15] Similarly, as Denmark had suffered a great economic and geographical loss following the Napoleonic Wars, the prospect of extracting resources made field science a high priority alongside the trade of natural resources. The surveys of Greenland were linked with Danish nation-building in other ways. As in the British context, cataloging the empire, knowing the land and the people, meant collecting natural history specimens. The Danish Crown requested that as many specimens as possible be sent to the Botanical Garden (Botanisk Have) and the Royal Museum (Kongelige Museum) in Copenhagen, and for use in projects that cataloged the natural history of the Danish Empire in the eighteenth and nineteenth centuries, such as *Flora Danica*.[16]

Travel narratives typically included images of the landscape, the flora and fauna, natural phenomena such as the aurora borealis, and Indigenous peoples. Visual representations of the Arctic have been the focus of recent scholarly literature.[17] Such studies show the significance of visual imagery in shaping conceptions of the Arctic as a space, and focus on both images in books and periodicals, as well as the large and popular Arctic panoramas that were on display throughout the nineteenth century. Arctic explorers, as well as those in other regions of the world, surveyed and mapped unknown lands, and visual imagery including maps played a key role in making the foreign tangible. I am particularly informed by the work of the historian Daniela Bleichmar, who has shown how scientific images reveal the intimate relationship between knowing and making visible within the Spanish imperial project. Bleichmar developed the concept of "visual epistemology," which emphasizes how observation was a trained and highly situated practice.[18] Drawing on these insights, I show throughout this book how illustrations, as well as highly visualized language, were tools through which Arctic travelers simultaneously affirmed their imperial rights to the land and themselves as authoritative observers. The process of making the Arctic visible was also a process of erasure. For many European and Euro-American explorers, rendering invisible the work of Indigenous peoples was part of their own strategies for positioning themselves as experts on the Arctic regions. I argue that this strategy of erasure, or rescripting of labor, formed part of what Jane Burbank and Frederick Cooper have termed the imperial "repertoires of power," for establishing and legitimizing imperial authority.[19] The case studies in the following chapters all show that when considered within their imperial context, travel narratives reveal significant and overarching geopolitical considerations. The way this was represented relied heavily on the framing of the explorer.

THE EXPLORER

Who was an authoritative observer of Arctic phenomena? Who was the Arctic explorer? In answering these questions, I take my starting point with the historian of science Janet Browne, who has identified three main categories of traveling naturalists and collectors:[20] freelance and independent entrepreneurs, navy or military employees, and those employed to collect natural history specimens. I also examine narratives from additional categories of Arctic explorers, including Indigenous peoples, missionaries, private entrepreneurs who relied on patronage, and those employed by trading companies. Although settlers such as missionaries were engaged in different activities than those employed on exploratory missions, they formed an influential part of the imperial projects in the Arctic. There is also the changing recombination of such categories, which serves to illustrate how narrative choices and their effectiveness were linked to the identity of explorers and organizing bodies. Uncertainty, and how it influenced the nature of the Arctic expedition, was highly situated. Factors such as the organizers, the national contexts, narrative style, as well as what the historian of anthropology Henrika Kuklick has called the "personal equations," were all part of shaping the practice and perception of nineteenth-century fieldwork. The identity of the travelers and their social circumstances were central to shaping the nature of expeditions, and as the authors of *Travels into Print* have observed, "Questions of epistemology and truth telling in print were ineluctably linked to the status of one's informant, the social standing of the author, or the warrant by association that came with being officially sanctioned to have undertaken the travel or the exploration by a government or a scientific body."[21]

The question of who is an explorer and what is a narrative points back to our understanding of travel. Notably, the historian of anthropology James Clifford has observed that the concept of travel is a complex range of experiences. He proposes that "to see fieldwork as a travel practice highlights embodied activities pursued in historically and politically defined places," thus emphasizing the intersecting routes and reconfigurations of borders (intellectual and spatial) for the traveling fieldworker. It is therefore not surprising that it becomes possible to expand our perception of who is an explorer when we reconsider what it means to travel. Yet this has not been consistently done for the Arctic context.[22] Especially when it comes to studies of scientific knowledge-making in the Arctic, the historiographical focus has overwhelmingly remained on specific European and Euro-American explorers. But Arctic Indigenous peoples working as part of expeditions played a key role in these ventures. They not only facilitated exploration and scientific research, but actively undertook this work as well. In addition, Arctic Indigenous peoples working as part of European and Euro-American ventures were often also strangers to the surveyed lands and the peoples living there. When we reconsider who is an explorer, it opens up new perspectives, both exciting and challenging, to otherwise well-known stories.

The situatedness of the perception of who was an authoritative observer of the Arctic is particularly evident in the "heroic Arctic explorer" trope.[23] Kuklick has further argued that perceptions of fieldwork and the associated physical ardor changed from dirty and ungentlemanly to heroic. This assumes that those working in the field were part of a lower social status than the gentlemen-scientists who made use of the collected data and specimens. However, this characterization is problematic, especially when applied to the fieldwork of Arctic explorers.[24] Although this type of fieldwork was arduous, the explorer was framed as a gentleman who overcame danger and adversity to command nature at his will. But perceptions of gentlemanliness, heroism, and expertise were fickle. The strategies that allowed explorers to portray themselves in this desired way could have unexpected consequences. One such strategy was references to direct observation. An appeal to firsthand experience as a way of generating credibility was utilized from the first expeditions following the Napoleonic Wars, but the role of fieldwork in the Arctic was complex. In the British context, the explorer was described in heroic terms from the beginning of the nineteenth century. It is important to further consider the differences in organization structure of expeditions. The HBC and the KGH primarily undertook overland surveying ventures, while the ones organized by the British Navy were both overland and nautical. The format typically adhered to in the expeditions organized by the British Admiralty consisted of a large crew and two vessels. The ships were furnished with material items to make the journey more comfortable and a large selection of expensive scientific equipment, all of which were part of the construct of the officers' gentlemanly, or heroic, status.

In the Danish context, and with the expeditions organized by the HBC, the explorer was a different sort of fieldworker. Smaller-scale sled expeditions required a different approach, one which drew heavily on the insights and assistance of Indigenous peoples. This was particularly the case with those expeditions that to a larger degree combined multiple modes of traveling where we see the distinction between Kuklick's fieldworker and gentleman break down. That is not to say that the differences in exploratory format are irrelevant—to the contrary, they are essential. Differences in the mode of exploration shaped everything relating to the ventures, including the portrayal of the explorer and the science they produced, which is why a comparison of such differences reveals the complex role of fieldwork in identity-making. While there is no single answer to the question of who is an Arctic explorer, it is clear that the style of exploration and the organizing bodies involved had a significant impact on expedition formats and the resulting framing of their experiences and discoveries. Writing a travel narrative was an opportunity to present your expedition in a desired way—and to gather support for future projects. The way you framed yourself before, during, and after the expeditions was central in shaping the long-term success of your venture. Language and symbolism formed, and was shaped by, the construction of personal identities and scientific research through the travel narratives.

NARRATIVES

Opening up the categories of exploration and travel literature to include many types of travelers and their accounts decenters the moment of discovery, or lack thereof. This allows us to bring to the fore key issues of authorship and the function and construction of scientific knowledge in the Arctic. As the historians Elizabeth Bohls and Ian Duncan have noted, "Travel writing as a form or genre is not easy to pin down." Just as nineteenth-century readers encountered the Arctic through multiple types of firsthand accounts, I allow for a broad range of difference in the stylistic and narrative structure of travel narratives. This challenges our notions of who constituted an authoritative Arctic writer and how this impacted the wider questions about veracity and the production of scientific knowledge in the Arctic. The authors of *Travels into Print* argue that "travel writing is an analytical and interpretative category whose study involves the textual and stylistic analysis of works of travel and of exploration and, particularly of authorship, the style of writing, its underlying purpose, and the power of such writing to delimit, explain, or misrepresent the objects of its attention." Drawing our attention to the complex composition of nineteenth-century travel narratives, it speaks to the broad influence of travel writing, in all its forms.[25]

Travel narratives were closely tied to concerns over imperial authority in the Arctic. I draw on the historian Mary Louise Pratt's seminal work, which shows how European travel literature from the extra-European context visualized and shaped relations and knowledge, and how the identity of the explorer influenced the choice of narrative. Similarly, the editors of *Politics, Identity, and Mobility in Travel Writing* have noted that "we could view travel narratives as renegotiating cultural boundaries even while they actively establish such boundaries."[26] This emphasizes how travel narratives, rather than simply accounting for a voyage, are inherently political: they are not only linked to obvious geopolitical issues, but also to individual politics. As was noted in *Fraser's Magazine* in 1853, "Good travel-writing requires a certain sort of egotism."[27] Career ambition, friendship, scientific competition, love—those were but some of the personal aspects that shaped the narratives. Issues of boundaries and politics emerge throughout this book, from the charting of the Arctic coastlines (a very physical boundary) to the choices of narrative format for the travel accounts (an intellectual boundary). There were also boundaries between truth and falsehood.

Although I am not concerned with distinguishing truth from falsehood, nineteenth-century readers of Arctic explorations were. The historian of science Steven Shapin's examination of what it means to present yourself as truthful in relation to the organization of science has been particularly influential in how I treat this issue. As Shapin wrote in *The Scientific Life*, he sought to "describe who truth-speakers are in late modernity: what kinds of people, with what kind of attributed and acted-upon characteristics, are the bearers of our most potent forms of knowledge." Drawing on this point, I show throughout the book that explorers sought, and often failed, to construct their narratives in such a way that their observations were perceived as credible. Perceptions of truthfulness were crucial. However, what constituted a trustworthy account was not straightforward, and the self-representation of Arctic explorers as authoritative and truthful observers of Arctic phenomena was not always effective. A key feature of travel writing as a genre was that authors read each other, repeating, commenting upon, and adjusting each other's points. This dialogue between the author and past explorers could work both to further or discredit cultural and scientific authority. It is useful to consider how in *Leviathan and the Air-Pump* (1985), Steven Shapin and Simon Schaffer developed the concept of virtual witnessing which "involves the production in a reader's mind of such an image of an experimental scene as obviates the necessity for either direct witness or replication."[28] As they show, choices regarding knowledge production and authority within a research field was linked with the self-portrayal of the natural philosopher as "objective" and "modest." Such considerations were also present for Arctic explorers.

Printed media were important sources through which information such as news, gossip, almanacs, and advertisements spread, including those related to the organization and results of expeditions. The representation of the Arctic in the nineteenth-century British periodical press has recently become the subject of scholarly interest, notably in the important work of historians and literary scholars such as Adriana Craciun, Jen Hill, and Janice Cavell.[29] This growing scholarship has drawn our attention to the many varied expressions of the Arctic project in print form, from elite nineteenth-century literature to the general periodical press. The British periodical press underwent significant transformations in the middle of the nineteenth century. It grew rapidly as new types of publications emerged. In Victorian England scientific news was of particular interest. Significantly, the periodical press provided a battleground for questions of authority, status, and cultural elitism in Victorian society. The transformations that took place in British science during the nineteenth century were rooted in a combination of several factors, such as the emergence of a growing reading audience, changes in paper taxation, developments in print technologies, and the telegraph.[30] As the editors of *Science in the Nineteenth-Century Periodical* noted, "From the perspective of readers, science was omnipresent, and general periodicals probably played a far greater role than books in shaping the public understanding of new scientific discoveries, theories and practices."[31] The British periodical press was highly significant in shaping knowledge and opinions about the Arctic and future Arctic expeditions. This draws our attention to the fact that news about Arctic voyages had circulated in the press prior to the publication of Arctic narratives and highlights the interplay that existed between book and periodical. Narratives were not constructed or read in a vacuum.

Drawing these perspectives together, I show throughout this book that the process of writing travel narratives was political, involved more figures than the listed author, and influenced the textual construction of Arctic science. This brings us to periodization and transnational comparisons. The historians of science Hans Henrik Hjermitslev and Casper Andersen have pointed to an important difference between the British and Danish contexts: in Britain the cheaper forms of printed materials, including popular science publications, appeared in the first half of the nineteenth century; in Denmark they did not appear until the second half of the century.[32] While Barrow was lobbying to organize another expedition, continental Europe was experiencing a period of unrest following the French July Revolution in 1830. Charles X was forced to abdicate, and uprisings throughout Europe, including in Poland, Italy, and Belgium, followed the July Revolution. In Denmark there was widespread dissatisfaction, as only around 2.8 percent of the population had the right to vote. King Frederik VI made some concessions to requests for democratization with the establishment of four Assemblies of the Estates of the Realm introduced by the laws of May 28, 1831, and May 15, 1834. The political restructuring in Denmark also extended to the border with Germany, the First Schleswig War between 1848 and 1851. This war concerned the area of southern Denmark and northern Germany called the Duchies of Schleswig and Holstein. While Denmark officially won the war, the issue was far from resolved, and it was reignited some fifteen years later with the Second Schleswig War.

Democratization and freedom of the press went hand in hand. While the context for scientific publishing was different in the national contexts, there is an important similarity: the increasing use of the periodical press as part of establishing scientific and cultural authority—in spite of war and restrictions on freedom of the press. Although the Danish Assemblies of the Estates of the Realm made it possible for journalists to discuss politics more critically, freedom of the press was still severely restricted under the Danish absolute monarchy, and Frederik VI imposed more and harsher restrictions, which led to the formation of the Society for the Proper Use of Freedom of the Press. The society played a key role in a series of reforms throughout the 1840s. The political unrest culminated on June 5, 1849, when the new constitution (*Grundlov*) established a constitutional monarchy. The establishment of the Danish constitution was in many ways a response to the 1848 Revolution in France where King Louis Philippe abdicated and the French Second Republic was founded.[33]

As in the British and Danish contexts, a range of factors influenced the growth of scientific and general publishing in nineteenth-century Canada, including changes in print technologies, rapid transatlantic and railway services, and increased literacy. While I focus on the legacies of British imperialism in Canada, the French context for Canada should not be forgotten, especially as we consider the plurality of imperial cultures and languages. The context for science and scientific publishing in Canada in the nineteenth century was shaped by the political turmoil of that period. While science was a popular topic in the periodical press in Britain, both in specialized journals and general newspapers, this was not the case in Canada. In the first decades of the nineteenth century there were only a handful of English-language newspapers in Canada, with numbers expanding rapidly in the 1840s and 1850s. Although there were hundreds of newspapers and specialized periodicals in print in the second half of the nineteenth century in Canada, only a few of these were dedicated to scientific topics.[34]

The differences in development of cheaper forms of printed materials and the general reader for science in each country shaped their publication and reception of travel narratives. Even within Western Europe there is no meaningful unified periodization of developments in print culture and science. It would therefore be a mistake to apply British concepts of a communications revolution to other countries, thus highlighting the usefulness of a comparative, transnational, analytical approach.

A TRANSNATIONAL PERSPECTIVE

Arctic explorations were inherently transnational in nature. Explorers from different nations read and commented upon each other's narratives, and expeditions often included assistants from other countries, in addition to Indigenous peoples hired in the Arctic. The Arctic is a vast polar region, currently considered to spread across Canada, the United States, Russia, Denmark, Sweden, Norway, Finland, Iceland, and the Arctic Ocean. European and Euro-American understandings of the Arctic changed throughout the nineteenth century. In this book I focus on Danish, British, and English-Canadian expeditions to Greenland and the Canadian Arctic, with references to select American expeditions. Although Denmark has a long historical presence in the Arctic and is one of the current eight Arctic states in the Arctic Council, there is a relatively small body of critical research on this history. I believe that a primary reason the Euro-American and British presence there has been considered of more historical significance than the Danish is due to the differences in exploratory and narrative strategies favored in each country. At the same time, while the Euro-American and British context has been thoroughly engaged with the literature, it has almost exclusively been in the shape of nation-focused accounts of Arctic exploration. Yet Arctic explorations were international projects, relying on the support of peoples and organizations from different nations, and they contributed to an international body of research in and about the Arctic.

The decision to focus this book on the Danish, British, and English-Canadian expeditions to Greenland and the Canadian Arctic is not unproblematic; for example, I do not engage with the French-Canadian or Russian historical contexts. My aim in this book is to show how the stories of Arctic exploration and scientific fieldwork, some of which are well known and others less so, take on new meanings when considered as part of their international context, rather than as national projects. There had been long-standing cooperation and scientific conversations between British, Danish, and English-Canadian explorers and settlers, and much of this took place in what is now Greenland and Arctic Canada. For example, British expeditions to the Arctic often relied on official Danish government support, as they harbored in and gained supplies from Danish settlements in Greenland. Because this contact was significant and continued throughout the nineteenth century, the Danish-British focus is a particularly useful backdrop to consider why specific explorers and expeditions have gained prominence in the retelling of Arctic history.

There is a large body of recent literature on transnational history that addresses the methodological advantages and difficulties of undertaking transnational research. Notably, the historians Michael Bravo and Sverker Sörlin have illustrated how limiting the study of scientific practices to one national context constricts our understanding of Arctic science. They observe that there was a difference in the northern narratives in Denmark and Sweden, as "the Danish approach was more spiritual, and spearheaded by missionaries, whereas in Sweden taxation, science, and even forced labor were the instruments. The northern narrative of Sweden, as a result of this, became much more concerned with resources and wealth, which was yet another similarity with the British imperial project."[35] Because of this difference in emphasis between Denmark and Sweden (and Britain, which they argue was similar to Sweden), there was also a difference in the perception of the Indigenous populations. The Danes held more positive attitudes toward Indigenous Greenlanders than Swedes, shaped by a paternalistic concern in combination with perceptions of guilt over their treatment. The transnational approach taken by Bravo and Sörlin is similar to that outlined in the American Historical Review's Conversation column "On Transnational History," which understands transnational history as a conceptual tool that allows historians to think differently—most importantly, to think about and follow movements, flows, and circulations of peoples, ideas, knowledge, and objects.[36] Compared with other types of historical methods, such as world history, transnational history multiplies the focus from the state to many types of actors moving across boundaries.

In writing about the many methods of studying transnational history, the historian Patricia Seed observed that "transnational history has multiplied the foci of research from the state alone to a variety of independent transnational economic actors—individuals, communities, migrants, or organizations that may have played independent roles in the economic growth of a city, state, or region."[37] Trading routes connected the oceanic spaces but exchanges were not limited to commercial goods or economic concerns. For example, the historian Sugata Bose has argued for a historical approach that moves beyond a focus on trade relations to elucidate the economic dimensions of interregional integration. By broadening the focus to include topics such as geopolitical, military, cultural, and religious issues, Bose sees oceans within their imperial context as an interregional arena. From this perspective, oceans become something between the local and the global, consisting of "a hundred horizons, not one, of many hues and colours."[38] Greenland and Denmark were connected by a steady flow of commercial goods, as well as ideas, experiences, and people. The character of this interconnectedness was more established in the case of Greenland and Denmark, as well as in the HBC territories, than among Britain and their explored Arctic territories. While missionaries and employees in the trading companies did not live permanently in the Arctic, they were parts of networks that had long-term settlements in the territories they explored. By contrast, the purpose of the British Royal Navy–sponsored expeditions was to explore and return home to Britain.

The extent to which the imperial metropole was able to control the results of expeditions in the periphery was limited. This was the case both in the Danish and British imperial contexts, though there were clear differences in how the organizing bodies attempted to lower the uncertainty of the Arctic as a field site for scientific research. The identity of explorers both influenced and was shaped by the imperial context of exploration. Although many historians have examined Arctic explorations, in particular those associated with John Franklin (1786–1847), there is still much to be gained by studying the scientific practices of

Arctic explorers and their repertoires for establishing knowledge claims in their narratives. I combine the four broad historiographical themes outlined above— the intersection of imperialism and science, exploration identities, studies of travel writing, and transnational historical methods—to shed new light on the function of travel narratives as scientific documents and the formation of field-based science in the nineteenth century within the nexus of imperial expansionism, international competition, and attempted cooperation in the icy north. This complicates our understanding of scientific research in the Arctic and the various sociopolitical factors that shaped that construction. Throughout the book, I compare and contrast the Danish, British, and Canadian presence in the Arctic, while also touching on the perceptions and attitudes toward international collaborations. In all cases I argue that a more comprehensive understanding of the Arctic as a field site can be developed through a transnational perspective on travel narratives and the identity of the Arctic explorer. In doing so, I offer a new way of looking at narratives, as not simply an account of a voyage, but as a way to unpack the inherent international and highly fraught nature of nineteenth-century Arctic exploration and scientific fieldwork.

STRUCTURE OF THE BOOK

The period between 1818 and 1883 was shaped by several key transitions in Arctic explorations. In order to avoid the temptation (or risk) of writing an exhaustive (or exhausting) account of all nineteenth-century Arctic expeditions, I have centered each chapter around a selection of Arctic explorations and their narratives. The disappearance of Franklin's expedition was a transformative event, but it was not the only one, and not necessarily the most significant one either. For this reason I do not conclude this study with the last official British expedition in search of the *Erebus* and *Terror*, but instead with the First IPY in 1882–1883. I examine the narratives from these expeditions, and depending on the one in question, discuss its publication and reception in both general and specialized periodicals as it relates to the construction and practice of science in the Arctic. In this way I have adopted an approach to studies of the nineteenth-century Arctic that can be described as fitting between those that focus more exclusively on the expeditions' scientific results and those that put the emphasis on the textual and visual representations of the Arctic.

Each chapter has three main case studies that are roughly chronologically organized. I have sought to balance the focus between the British, English-speaking

Canadian, and Danish contexts, and I have chosen expeditions and explorers that provide a certain level of thematic continuity across an otherwise diverse set of examples. My aim has been to trace similarities and differences in scientific practices, attitudes toward exploration and colonial expansion, and the ways scientific knowledge was communicated in multiple national contexts. I point out four major transitions: The theme of Chapter 1 is beginnings, but it could also have been "uncertainty." The radical uncertainty of the early expeditions extended to the Arctic explorers themselves, as narrative strategies for establishing scientific and cultural authority through the travel accounts were negotiated. The theme of Chapter 2 is economics, where I consider the interconnectedness of commercial goods, ideas, experiences, and people, and examine the way the tensions over financial gain impacted the nature of Arctic explorations and the perceptions of the Arctic explorer. Opportunism is the theme of Chapter 3, reflecting the economist Oliver Williamson's famous description of opportunism as "self-interest seeking with guile." His discussion of opportunism and economic actors is similar to the "Opportunism-in-Context" Model developed by the philosopher Andrew Pickering which draws attention to how researchers made use of their available resources in different contexts.[39] This emphasis on the role of opportunism can usefully be extended to Arctic exploration in the post-Franklin era. With the disappearance of John Franklin's expedition, the number of expeditions multiplied. Searching for his expedition was the opportunity, but the goal was, as before, intertwined with economy, glory, and power. The many search missions were followed by Arctic exploration fatigue in Britain, while other nations began to stamp their authority on the region. The theme of Chapter 4 is therefore internationalism, as I show how the transformations in imperial authority and attempts at international collaboration with the First IPY challenged old perceptions of the Arctic explorer and scientific research in the Arctic.

Chapter 1 shows the disunity of Arctic science in the early part of the nineteenth century, bringing out the discord between the desires of figures in the metropole and the reality of explorations in the High North. I focus on the 1818 expedition led by John Ross, William August Graah's voyage to the east coast of Greenland between 1828 and 1829, and John Franklin's Coppermine expedition between 1819 and 1822. In this period British expeditions largely focused on discovering a Northwest Passage, while a key aim of the Danish expeditions was to establish evidence for a historical Danish presence in Greenland to support its often-disputed territorial claims to the region. Despite this difference, the expeditions had a central, overlapping feature: a discrepancy between the originally stated aims of the expeditions and what they actually achieved. While figures such as John Barrow played a key role in determining the makeup of the British voyages and the career trajectory of the British explorers, there were limitations to this control from the metropole. This was also the case in the Danish context. The chapter shows that the nature of scientific research in the Arctic in the early years following the Napoleonic Wars both created and was shaped by the uncertainty associated with Arctic expeditions, the unstable nature of intellectual and cultural authority, choices of narrative styles in the travel literature, encounters with Indigenous populations, and the persona of the Arctic explorer.

In the British context, the disillusionment with the search for the Northwest Passage opened up opportunities for other players to take center stage. Lack of funds created a similar situation in the Danish context. Chapter 2 looks at four expeditions that were funded and organized in the 1830s outside the realm of government: John Ross's second and last expedition to the Arctic between 1829 and 1833, Peter Warren Dease and Thomas Simpson's expedition organized by the Hudson's Bay Company between 1836 and 1839, and accounts by the Danish pastor Johan Christian Wilhelm Funch and an anonymous Danish woman missionary, both in Greenland in the 1830s. A key theme is the ambivalent relationship among religion, commerce, and science, and how this influenced the prioritization of formal scientific inquiry and the use of expensive equipment such as chronometers. The chapter shows that there was tension between the types of scientific results that were expected from exploratory missions and the focus of the trading companies and religious missions.

By 1844, after numerous failed attempts, the second secretary to the British Admiralty, John Barrow, was eager to promote one last expedition in search of the Northwest Passage. John Franklin volunteered, and he left England with his crew aboard the *Erebus* and *Terror* in 1845. The disappearance of Franklin's expedition changed the context for subsequent Arctic expeditions. The vagueness of the goal, finding the lost Franklin expedition, allowed for more flexibility in terms of what activities could be conducted during these search missions. Their official goal generated more opportunities for Arctic exploration. Yet it appears that this was not always the primary motivator behind them. In Chapter 3 I interrogate the nature of Arctic science when carried out under the added pressure of finding the Franklin expedition, with a focus on John Rae and John Richardson's expedition between 1848 and 1849, Rae's later report that Franklin's men had resorted to cannibalism, and Carl Petersen's participation in the 1857 search under the command of Francis Leopold McClintock. The change in primary goals challenged previously held conventions for what Arctic expeditions should accomplish. A key theme throughout this chapter is the stark differences in the reaction and response to Franklin's expedition between Denmark and Britain and how this influenced the production and representation of scientific research during the expeditions.

The period between McClintock's expedition and the First IPY was characterized by a transition in colonial power in the Arctic, which influenced all aspects of how Arctic expeditions were carried out, from their style to their interactions with Indigenous populations. Chapter 4 shows how nationalistic concerns were also linked to apprehensions about changes in the Arctic field site and the identity of the Arctic explorer. I examine these shifts through the Indigenous Greenlandic explorer Suersaq's participation in George Nares's expedition between 1875 and 1876, the establishment of the journal *Meddelelser om Grønland* in 1879 (with a focus on the expeditions led by Knud Johannes Vogelius Steenstrup and Jens Arnold Diderich Jensen in the late 1870s), and the First IPY in 1882–1883. My main focus for the IPY is on the British-Canadian contribution—the polar station at Fort Rae—in comparison with the American contributions to highlight the relationship between changes in Arctic fieldwork and narrative practices. In this way the chapter brings to the fore the connections between the cautious international cooperation in this period of transition of imperial authority in the Arctic, changes in scientific practice, and the identity of the Arctic explorer.

Altogether this book is concerned with questions about what constituted scientific research, who were considered scientific practitioners, how this vast area that we today understand as the North American and Greenlandic Arctic was understood in the nineteenth century, and the way this knowledge and definitions changed in time and place. By approaching surveying in its broadest sense—as the ordering and quantifying of nature through travel as a way to conceptualize the scientific practices of the Arctic explorers—it is the aim of this book to show how abstract notions about the Arctic became tangible in the nineteenth century.

CHAPTER 1

New Beginnings in the Arctic

The expeditions which have recently been engaged in for discovering a North-west passage, though unsuccessful in their main objective, are generally, and very properly, considered undertakings of great utility. Conducted as such expeditions now are, they cannot fail of procuring many valuable additions to the arts and sciences; whilst the spirit of enterprise kept alive by them, both in officers and seamen, renders them an appropriate service in time of peace, for the employment of a small portion of that navy, which during the war established our right to the uninterrupted navigation of all "the mighty waters."

— Thomas Merton (pseud.), "Arctic Natural History," *Literary Magnet,* 1824

Following the end of the Revolutionary and Napoleonic Wars that took place between 1792 and 1815, there was a significant renewed interest in the Arctic. The possibility of discovering a trading route to the Pacific was a major incentive, but it was not the only motivator. Although geographical discovery was the official primary goal, scientific discoveries were central to the expeditions. This was especially so when faced with a lack of geographical results. The first expeditions organized by the British government following the wars were important in showing what could be accomplished scientifically in the Arctic. We can look to the expeditions' official orders for the types of science that sponsoring parties such as the British Navy valued. These included requests for experiments on magnetism, the aurora borealis, the figure of the Earth, refraction, ocean currents, mineralogy, zoology, botany, hydrography, ethnology, and the general collection of natural history specimens; all in all, such requests were very vague and very broad. Geography was also considered a science, but geographical discovery and mapping were evaluated separately by organizers and explorers. This vagueness of the official instructions to the expeditions is a clear indication of the difficulties faced by explorers before, during, and following their journey to the Arctic.

This chapter explores this uncertainty through three early nineteenth-century expeditions: the twin 1818 British Arctic voyages in search of the Northwest Passage and the North Pole led by John Ross (1777–1856) and David Buchan (1780–1838), with a focus on Ross; John Franklin's Coppermine expedition; and the Danish expedition to the east coast of Greenland under William August Graah. In contrast with Ross's voyage, both Graah's and Franklin's expeditions relied heavily on the assistance of trading companies and Indigenous communities, and the trajectories of their projects were shaped by this support. In all three cases the results were wide-ranging and largely dependent on the abilities of the crew, as well as the luck of the expeditions, the environment, and the people they encountered before, during, and after the expeditions. Consequently, there was not always a match between the expeditions' goals, scientifically and geographically, and what they actually produced.

That is not to say that people such as the second secretary of the Admiralty John Barrow did not play a key role in determining the makeup of the voyages and the career trajectory of the explorers, but there were clear limitations to their control. There was no unified practice of science in the Arctic, and no functioning blueprint for creating a successful expedition. Just as the experiences of the crew and what they were able to achieve were markedly unpredictable, what followed upon their return was also hard to control. As the case of Ross illustrates, the textual strategies employed to navigate the expectations of expeditions were difficult to manage and formed a central part in the quick downfall of his career and public persona. This had to do with the unstable nature of intellectual and cultural authority. Both the variability and perception of the results were shaped by the stylistic choices in the travel narratives, including the construction of the Arctic explorer's persona. In the early nineteenth century the nature of scientific practices in the Arctic and the publication of travel narratives were shaped by uncertainty and instability. It was not possible for the metropole to determine the results from and reception of the expeditions.

A VOYAGE OF DISCOVERY: THE FIRST BRITISH ARCTIC EXPLORATIONS

During a voyage to the Arctic in 1817, the whaling captain and naturalist William Scoresby (1789–1857) noticed that there was less polar ice than usual. A decrease in polar ice meant an increase in open water, and a potential for reaching areas which had hitherto been impenetrable by ship. Scoresby informed the influential

British naturalist Joseph Banks (1743–1820) of this change. He suggested to Banks that it would be an opportune time for the British government to sponsor an expedition—although he warned that the amount of ice might not stay stable for the following season. Scoresby had his own reasons for bringing this to the attention of the government: he wished to lead the suggested expedition. Intrigued by Scoresby's information, Banks in turn counseled Robert Dundas (1771–1851), also known as Lord Melville, the first lord of the Admiralty, on the possibility and opportunities for discovering a Northwest Passage. Dundas, as well as Barrow, were quick to see the possibilities offered by this apparent change in the Arctic climate, but it was not Scoresby they had in mind to actualize their project.[1]

Another factor in making the decision to organize the first Arctic expeditions after the wars was the changing geopolitical situation of the early nineteenth century. During the wars British naval science and Arctic explorations had been on hold. In 1812, 113,000 seamen had been funded by the British Parliament, but this fell to 24,000 in 1816. Up to 90 percent of officers were unemployed by 1817, and peacetime meant that this large number of un- and underemployed seamen were keen to secure a post.[2] The expansion of the Ordnance Survey provided a key opportunity for employment for these men. Officers trained in scientific surveying were useful in this revival of British Arctic exploration. Even more so, their skills and approach to fieldwork played a big part in shaping efforts to implement the grand geopolitical dreams of establishing authority in a region which seemed to promise a faster route to the Pacific.

Barrow had several motives for supporting Arctic explorations. The economic possibilities from a potentially faster and safer trading route were obvious, but national pride was also a significant factor. The advance of science and the national glory associated with such scientific progress factored heavily in Barrow's thinking. And in the event the Northwest Passage was not discovered, Barrow hoped that there would be significant scientific discoveries resulting from the expeditions. These could, for example, be utilized to advance knowledge of the British climate to improve agricultural practices. The use of foreign land as a laboratory for the advance of British science has a long history. For example, the historians Richard Grove, Deborah Neill, Nancy Stepan, and Katherine Anderson have shown how British colonialism in the tropics facilitated studies that extrapolated from a specific environment to new evaluations of (British) nature more generally.[3] While the primary goal of Barrow's push for Arctic exploration was geographical, the secondary purpose was scientific. This tension—between the very clear aim of finding a passageway and a more general idea of scientific advancement—mattered greatly at all stages of the Arctic exploration, starting with the choice of vessels and crew.

The people selected to lead the twin 1818 voyages in search of the Northwest Passage and the North Pole were John Ross and David Buchan. Buchan and his second-in-command, Lieutenant John Franklin, had command of the ships HMS *Dorothea* and HMS *Trent*. The goal of Buchan's expedition was to reach the Pacific Ocean through the hypothesized Open Polar Sea and the North Pole. Ross's expedition sought to reach the Pacific Ocean by going through Baffin Bay and Lancaster Sound. Ross's second-in-command was Lieutenant William Parry (1790–1855), and together they had command of the HMS *Isabella* and HMS *Alexander*. As the first voyages to the Arctic after the Napoleonic Wars, Ross's and Buchan's expeditions were central in establishing British dominance in the Arctic region and in showing what could be accomplished with future ventures to the Arctic. The problem was that neither of these expeditions were particularly successful, if success was to be judged from their goal of finding a Northwest Passage.

Ross accounted for his expedition in *A Voyage of Discovery, Made under the Orders of the Admiralty, in His Majesty's Ships Isabella and Alexander, for the Purpose of Exploring Baffin's Bay, and Inquiring into the Probability of a North-West Passage* (1819).[4] Ross's narrative was influential in shaping perceptions of the Artic and what did, and did not, constitute an authoritative Arctic explorer. As Ross would learn the hard way, the techniques used to establish an authoritative voice shaped the tone of the narrative. This in turn affected the description of the Arctic, the science carried out there, and the nature of the Arctic explorer. Originally the intention was to publish an official account, sanctioned by the head of the British Navy, of Ross's and Buchan's voyages. In both cases, this was decided against, in part because the expeditions had come nowhere close to fulfilling their geographical goals. Ross published his, now unofficial account of the expedition in 1819. Buchan, on the other hand, did not publish an account of his attempt to reach the North Pole, due to what he perceived as their lack of results. A full narrative of Buchan's voyage was not made until Frederich William Beechey (1796–1856), a lieutenant on the expedition, published *A Voyage of Discovery towards the North Pole: Performed in His Majesty's Ships Dorothea and Trent, under the Command of Captain David Buchan, R.N.; 1818; to which Is Added, a Summary of All the Early Attempts to Reach the Pacific by Way of the Pole* in 1843.[5]

As outlined in the official instructions to the expeditions, the primary focus of Ross and Buchan's voyages was geographical. The secondary focus was scientific, and the importance placed on this aspect is evident in the number of valuable scientific instruments and books on board the ships, as well as the detailed requests for a wide range of scientific results. In addition to playing a central role in deciding which officers were part of the Arctic expeditions, Barrow also determined which scientific instruments the Admiralty would purchase for them. It is for reasons such as this that Barrow's biographer Christopher Lloyd has described him as a figure who always appeared in the background directing the course of naval policy. Ross's conflicts with Barrow started early on, as Barrow denied his request for an additional timekeeper to bring on board the *Isabella*.[6] In the end Ross purchased the timekeeper himself. Barrow's influence did not end with the expedition, but extended through the post-expedition narration of the voyages. John Murray was an official publishing house for the Admiralty, and it produced a large part of the travel accounts from expeditions to the Arctic (and Africa) in the first half of the nineteenth century. Barrow functioned both as a preprint reader and post-publication reviewer for Murray. In this way he influenced both the physical makeup of the expeditions, their orders and instructions, and the portrayal of the expeditions and the Arctic upon the conclusion of each expedition.[7]

The key scientific areas of interest as outlined in their official instructions were to examine the variation and inclination of the magnetic needle, the intensity and variation of the magnetic force, the temperature of the air and of the surface of the sea, the dip of the horizon compared over fields of ice and open horizon, refraction of objects over ice, the character of the tides and currents, the depth and soundings of the sea, and the sea bottom. In addition to observations linked to meteorology, they were also to collect and preserve animal, mineral and vegetable specimens, and make drawings and descriptions of those they could not preserve and store on board the ships. The vagueness of this particular part of the official instructions illustrates the extent to which they were venturing into the unknown. It was impossible to know what could be expected and what could be discovered when (or rather, if) they reached their destinations. For example, the hope was that Buchan's expedition would reach the North Pole and there be able to make "the observations which it is to be expected your interesting and unexampled situation may furnish you with."[8] In Ross's official instructions, the Admiralty lamented their inability to provide detailed guidance with regard to route and time frame for the voyage, as the land in the region was unknown. Therefore, the instructions specified that they relied on Ross's skill and zeal for the safe fulfillment of the objectives of the voyage.

It is worth reiterating that while the primary goal of their expeditions were geographical, the official secondary purpose was scientific advancement. Many of the instruments the ships carried were related to one of the key research areas of Arctic explorations: magnetism. Terrestrial magnetism was a research area full of unknowns, one that greatly affected the practical aspects of seafaring. Knowing where you were was a central part of exploration. Since the time of the early astronomers, it had been possible to measure latitude with a fair amount of accuracy. Determining longitude, however, was more problematic. Whereas latitude, the position on a north–south axis, could be determined with the aid of stars or the sun, this was not enough to determine longitude, the position on an east–west axis. A fixed reference point of known longitude was needed, from which the ship's position at sea could be determined by the difference in time between it and the known position. Many European countries were interested in solving this issue. The British government established the Board of Longitude in 1714, with a prize for discovering a way to reliably determine longitude at sea. The Longitude Prize was awarded to John Harrison in 1773 for the design of a chronometer, a device that could keep time for months and was not easily affected by the conditions at sea.[9]

Another tool for navigation was the dip circle or dipping needle. The dip of the needle was recorded as part of studying geomagnetism. The needle moved in a vertical plane, to measure vertical magnetic inclination. At the Magnetic North Pole the needle in the instrument would point downward, as the magnetic field became more vertical. If Buchan's expedition had discovered the Magnetic North Pole, they would have measured a dip of 90°. Buchan's expedition came nowhere near either the Geographic or Magnetic North Pole, but both Buchan's and Ross's expeditions carried with them several dipping needles, as well as chronometers and magnetic compasses. The *Isabella*, Ross's ship, contained seven chronometers, three of which were the property of the British government, with four privately owned. (The *Alexander* had three government chronometers.) The *Isabella* further brought with it four dipping needles and seven compasses. They included multiple versions of the same type of instrument from different manufacturers to maximize the data they could produce and ensure their reliability.

▶ **FIGURE 1.1.** Illustration showing the method by which John Ross's expedition took transit bearings using a mark on the shore as a set reference point. *Source*: John Ross, *A Voyage of Discovery*, 1819, xvii. Courtesy of the Scott Polar Research Institute.

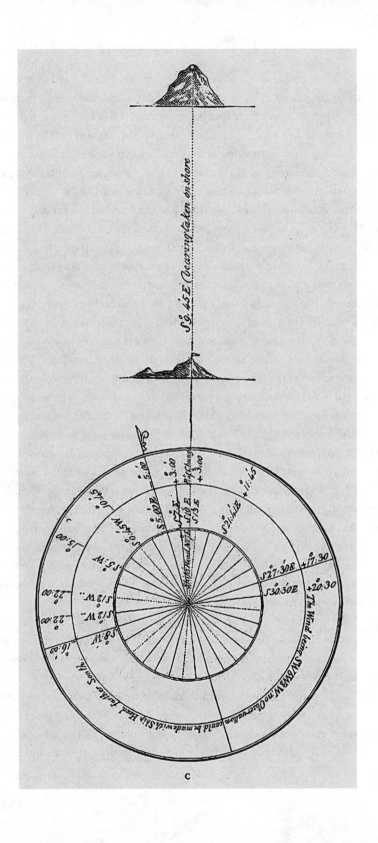

c

In order to obtain the true variation by reference to a fixed spot on land, the expedition placed a flag on a high point from which they used a compass to ascertain the bearing of a fixed spot on a mountain. The ship was steered to bring the flag and the fixed spot in one line, which allowed them to take the transit bearings. This process provided a reference for navigation with the compass, to add or subtract the degrees and minutes from the variation observed. Experiments and measurements such as these were included in the narrative, with tables and illustrations. We know from Ross's narrative and private notebook that they typically took the same measurements multiple times with different instruments to determine if any of the instruments provided outlying data. An average could then be calculated from multiple experiments. For example, during a quiet stretch of days at sea, Ross's expedition party made several observations with their instruments and found that Jennings's insulated compass was the median between them. One of the proposed disturbances was the iron in the ships. Because of the difference in results, Ross noted the name of each instrument and the person who had performed the observations in his notebook. By making multiple experiments with several instruments and from different points, Ross attempted to maximize the impact and accuracy of his scientific results. It was a way to eliminate mistakes, something that was later repeated in variation in the methodology of the First IPY as discussed in Chapter 4. The appendix of the published narrative also contained a "Report on Compasses, Instruments," and "Reports on Various Instruments Supplied to His Majesty's Ships Isabella and Alexander" that evaluated them. It provided a way for the instrument manufacturers to test if their devices were stable and accurate. In this way Arctic voyages functioned as a practical test spaces. The measurements made during Arctic explorations were key evidentiary sources for many research fields, even long after the voyage had taken place. Consequently, explorers faced multifaceted challenges when they wanted to construct narratives or publish scientific results and observations from their voyage.

Having taken part in a voyage was the first step in establishing authority, but visiting the Arctic was not enough in itself to create an authoritative voice. The explorer had to be considered a trustworthy observer and conveyer of scientific knowledge. A central part of the strategies for establishing authority in the Arctic travel narratives was the use of an active present-tense narrative voice, or what has been termed the "syntax of agency."[10] The captain of the Arctic voyages authored a narrative of the expedition, which could be published as an official account sponsored by the government or as a private publication.

FIGURE 1.2. Illustration of the shoreline surveyed during John Ross's expedition, showing the Croker Mountains (*center*). It was included in color in Ross's narrative. *Source*: John Ross, *A Voyage of Discovery*, 1819, 174–75. Courtesy of the Scott Polar Research Institute.

While such an account appeared to have a single author, it was actually a joint effort produced by the officers who were part of the voyage. Ross's unpublished notebook from the *Isabella* included several instances where the name of the person who had made an observation was mentioned—including James Clark Ross (1800–1862), Edward Sabine (1788–1883), and Ross himself. But while Ross made use of others', his narrative was framed in a language that emphasized his own direct observations. The focus was on his earnest reporting of everything he saw, nothing more and nothing less: "My nautical education has taught me to act and not to question; to obey orders as far as possible, not to discuss probabilities, or to examine philosophical or unphilosophical questions. . . . I have here attempted nothing beyond the journal of a seaman."[11] This style of writing is similar to what the historians Simon Schaffer and Steven Shapin have termed "virtual witnessing."[12] The narrative's journal format invited the reader to experience the voyage together with Ross, to see the Arctic through his eyes. As a rhetorical tool, virtual witnessing and emphasis on directly observable data were in frequent use in nineteenth-century accounts of nature. Added to this were the images Ross included in his narrative, such as the view of what Ross termed the "Croker Mountains" in Lancaster Sound. The mountain range included in the illustration was not observed by the other officers, and its inclusion as a key discovery relied solely on Ross's observation. The visual aids functioned as further support for his written description of the landscape there, to solidify his claim that there was no access through the sound. The portrayal of travel narratives as conveyors of matters of fact through a syntax of agency stands in sharp contrast to how the narratives were actually constructed.

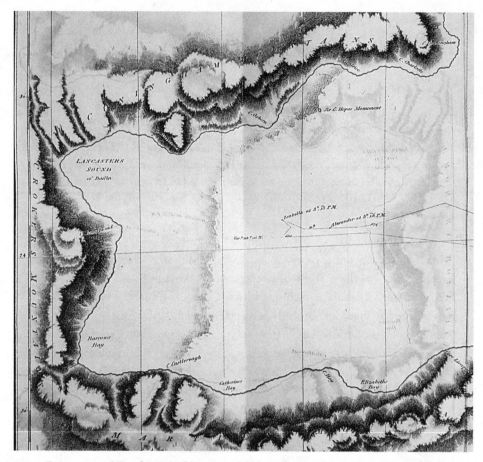

FIGURE 1.3. Map showing the route taken by John Ross's expedition through Lancaster Sound. The Croker Mountains are included to show that there was no passageway through the sound. *Source*: John Ross, *A Voyage of Discovery*, 1819, 174–75. Courtesy of the Scott Polar Research Institute.

The reception of Ross's narrative shows the complex nature of Arctic science, and the importance placed on scientific observations and experiments. As was remarked by an anonymous author in *Blackwood's Edinburgh Magazine*, "Few scientific enterprises in modern times have excited a more intense and general interest than those lately undertaken to the Arctic regions."[13] This aligns with Ross's emphasis on how little was known about the area, what explorers could expect, and how much could be gained scientifically from these expeditions. But the nature of undertaking scientific experiments and making observations during the voyages could cause problems. As part of his narrative, Ross had introduced the other participants of the expedition to the reader, including their roles and notable discoveries and contributions to the aims of the voyage. These descriptions contributed to his controversy with Sabine.

Sabine was described as the naturalist on board the *Isabella*, commissioned by the Royal Navy as help for Ross in achieving the scientific objectives of the voyage. But Sabine claimed that he had been unaware that this was his role. Rather, he argued that he had been asked to focus on variation and inclination of the magnetic needle, intensity of magnetic force, refraction, the aurora borealis, and the figure of the Earth, and not to collect specimens or make notes about mineralogy, zoology, or botany. In several instances Ross lamented that Sabine's qualifications had not been such that he could properly assist Ross. For example, Ross wrote that "with respect to the geology of this country, it is impossible to do more than to offer some conjectures, our Naturalist being unfortunately unacquainted with this subject."[14] This was not a very flattering portrayal of Sabine's abilities, and he was so unhappy with Ross's account of their voyage that he published a response, *Remarks on the Account of the Late Voyage of Discovery to Baffin's Bay Published by J. Ross, R.N.* (1819), the same year. Ross replied with *An Explanation of Captain Sabine's Remarks on the Late Voyage of Discovery to Baffin's Bay* (1819).[15] Sabine's complaints were threefold. First, the primary objective of Ross's voyage had been to ascertain whether there was a passage into the polar ocean from Baffin's Bay. In his narrative Ross made use of Sabine's statements in support of the conclusion that there was no passage through Baffin Bay via Lancaster Sound, but Sabine contested this use of his words. Second, Ross described Sabine as the naturalist on the voyage, but Sabine rejected this description. Finally, there was the issue of intellectual ownership of the scientific observations and measurements performed during the expedition. These complaints were closely linked to the way Ross had constructed himself as an authoritative observer.

Whether Sabine was officially the naturalist of the voyage extended beyond Ross's insults. If Sabine had been officially commissioned as a naturalist of the expedition, it would not be surprising that Ross as the captain and commander of the expedition made use of his observations, experiments, and measurements in an official narrative of the voyage—if the Admiralty had decided to publish one. When plans for such a narrative were abandoned, Sabine wrote to Ross requesting a return of his papers and informed him that he was considering publishing an account of the voyage. While Ross returned Sabine's papers, Ross's own narrative had, Sabine argued, made use of his work without his knowledge or permission, and additionally attributed credit for measurements and observations performed by Sabine to the captain's nephew, James Clark Ross, a midshipman on the *Isabella*. James testified in the presence of officers of the Royal Navy that he had copied the meteorological register from Sabine's personal notebook and provided them to his uncle. Part of his testimony was published in Sabine's pamphlet. Sabine questioned, "Did you not, when at or near Shetland, on our return home, copy my meteorological register for Captain Ross, at his request, and by my permission; being the same register that is engraved in plates in Captain Ross's book, and which was the only one so kept in the Isabella?," to which James responded, "Yes, I did." Sabine further charged that John Ross had reproduced his notes on magnetic observations and Inuktitut in an incomplete and incorrect form, stemming from his inability to read Sabine's handwriting. To counter these accusations, Ross claimed in his *Explanation* that his nephew had been misled by Sabine to believe that his was the only meteorological register kept on the ship. This was not the case, Ross argued, as the experiments and observations had been performed multiple times by multiple officers. His nephew had not seen the published data and had at the time of the interview been unaware that this and the data he had copied from Sabine were not the same.[16]

James Clark Ross wrote a letter to his uncle on April 13, 1819 wherein he recounted the interview, noting that he was "not conscious of a single point in which I have said anything to your prejudice." He explained how they had questioned him on specific observations, and that he "should not have been sorry" if they had asked more, as it would have given him the opportunity to account how his uncle had been present at a "great many" of the experiments, "but as to those on the Dip and Force I hope that when the Admiralty see clearly that you took one set & Captain Sabine the other, that all offences will be settled."[17] John Ross claimed that this plurality of registers and observers explained the discrepancies between Sabine's notes and the published works in his narrative. Different people were behind the data in the notes and in the published narrative.

Was Ross's textual strategy effective? In the immediate period after the publication of his narrative, the answer was both yes and no. The results of the voyage and Ross's narrative were widely discussed in the periodical press. It is clear that the relationship between Barrow and Ross had an impact on the general presentation of Ross's narrative, as well as later expeditions in search of a Northwest Passage. As the historian Janice Cavell shows, Barrow's insistence that voyages to unknown areas of the Artic were possible because of the change in ice, as Scoresby had reported, was contested by many.[18] The period between the return of Ross's expedition in 1818 and the arrival of news of Parry's successful discovery of a passage through Baffin Bay in 1820 was filled with uncertainty, where both Ross and Barrow could have been right at different times. Ross was a well-respected naval officer but Sabine's accusations had cast a serious cloud of uncertainty over the trustworthiness of his words. As a firsthand observer, Ross thought his statement that there was no passage through Lancaster Sound should have been trusted. The structure of his narrative was such that his statements were intended as a faithful description of the Arctic, with the active voice inviting the reader to experience the Arctic with him. The doubt raised by Sabine's accusations was a serious blow to Ross's credibility.

The uncertainty surrounding the trustworthiness of Ross was captured in the *Literary Gazette*, which on the question of Lancaster Sound noted that "we confess that we are against him in this hypothesis: he may be correct, but he certainly has not solved the problem. The very sound . . . which was most investigated, seems to be left in as much doubt as those Straits which were passed without examination."[19] The immediate reactions to the narrative were not wholly negative, and Ross was commended for ensuring the safety and health of his men, as well as for the quality of the scientific observations carried out during the expedition. For example, *Imperial Magazine* gave Ross a backhanded compliment: "We cannot, as some have done, pronounce this undertaking to have been altogether useless, though it has been ineffectual as to the attainment of its principal object." If the scientific results from the voyage had not made up for the lack of geographical advancement, at least they were a positive addition to knowledge: "The experiments made on the magnetic influence, and on the vibrations of the pendulum, the meteorological observations, the geographical determinations, and the discovery of a new Esquimaux tribe, that will undoubtedly be of essential service to future investigators, form a considerable accession to our stock of science and knowledge."[20] Many of these results were due to Sabine's work.

About Ross's style of writing, the *British Review and London Critical Journal* described it as exhibiting a "very praiseworthy modesty." Similarly, a review published in the *Edinburgh Review*, attributed to Murray Hugh (1779–1846), noted that "Captain Ross appears to have done his duty with great diligence, courage and ability; and to have told his story very clearly and honestly." However, the narrative itself, the review noted, was dull and heavy. The *Literary Gazette* referred to the situation with Sabine as a "misunderstanding" which was why the geological and natural history side of the scientific experiments were not as thoroughly carried out as would have been hoped.[21] The *British Review and London Critical Journal* charged the official instructions with being too vague, and "the whole of this code of instructions bears a crude and unphilosophical form, and reflects very little credit on the composer." The review further scolded Sabine and Ross for letting their personal affairs negatively influence the production of results: "Nor does there seem to be any direct and satisfactory way of accounting for the very great deficiency of detail in the department to which Captain Sabine's exertions were directed. Collisions of personal claims and private competitions are always at work to oppose the success of public undertakings, even on subjects of the most general interest to humanity." A detailed examination of Sabine's *Remarks* and Ross's *Voyage of Discovery* was printed in the *Edinburgh Review*. Here it was lamented that Ross had altogether made use of Sabine's words in this context; although they did not doubt the veracity of Ross's words, it cast a shadow of doubt on his account that extended to the rest of the book. While the *Edinburgh Review* noted that "we are of the number of those who regard a north-west passage from the bay of Baffin into the Pacific as a mere fancy never to be realized," it pointed out that a key problem was how Sabine's protest influenced Ross's perceived credibility.[22] This illustrates the work that went into creating trust in the data and how perceptions of credibility could be dismantled. The veracity of Ross's narrative itself had been questioned.

The whole situation was catastrophic for Ross and his career. By contrast, Sabine continued to enjoy a successful scientific career and participated in further explorations. The Board of Longitude was abolished in 1828 and replaced with the Resident Committee for Scientific Advice for the Admiralty. One of the three key figures involved in the new committee was Sabine; the two others were Thomas Young (1773–1829) and Michael Faraday (1791–1867). In 1826, just two years before the abolishment of the Board of Longitude, Sabine was awarded £1,000 for his pendulum experiments.[23] James Clark Ross also had great success, later accompanying Parry on his expeditions. Following the successful exploration of Lancaster Sound with the ships HMS *Hecla* and *Griper* in 1819, Parry led an additional two expeditions to the Arctic. Between 1821 and 1823 he was in charge of an expedition in search of the Northwest Passage on board the HMS *Hecla* and *Fury*. On this expedition Parry went from the Hudson Bay via Frozen Strait to Repulse Bay, where he found no further passage. They surveyed the coastline up the Gulf of Boothia and Baffin Island and found and named the Fury and Hecla Strait, through which was also no entrance into a Northwest

Passage. The third expedition between 1824 and 1825 took Parry to Prince Regent Inlet, where he believed the entrance to the Northwest Passage was located. Because there was more ice than they had expected, the party wintered in Prince Regent Inlet in Port Bowen. The *Fury* was damaged badly and abandoned. Parry left stores from the *Fury* on Fury Beach, and these reserves were later taken advantage of by other explorers, such as Ross in 1829 and McClintock in 1859. In addition to these three missions in search of a Northwest Passage, James Clark Ross also went with Parry toward the North Pole in 1827, but reached only 82°45'.

John Ross was never asked to return to the Arctic by the Royal Navy, and he spent the next several years ruminating over what had happened. It was a perfect trifecta of problems for him: he did not find a passage through Baffin Bay, he was accused of plagiarism, and then Parry showed that the "Croker Mountains" had been a mirage. The plurality of hidden authors of Ross's narrative had given the text the character of a frank, straightforward description of the Arctic from Ross's perspective, but this technique backfired, and Ross had to reveal the composite nature of his scientific observations to avoid the more serious charge of plagiarism. He was forced to show what was behind the curtain. As one of the first published narratives from Arctic expeditions in Britain following the Napoleonic Wars, Ross's difficulties illustrate how the uncertainties associated with Arctic research pertained not only to what to expect while in the region but also after the return home, as different people attempted to control the post-expedition narratives. As the example of Ross and Sabine shows, the techniques utilized to establish textual authority could also cause problems for the author. The strategies that Ross used to establish an authoritative narrative voice were what Sabine utilized to discredit him. Both had firsthand experience in the Arctic; they were both gentlemen and men of science. The unknowns of the Arctic and the uncertainty of what could be accomplished there scientifically meant that the construct of the Arctic explorer as trustworthy observer was a central part of the travel narratives, not just for the reception of scientific results but also for the career and social, cultural, and scientific status of the explorer.

THE HBC AND THE BRITISH ROYAL NAVY JOINS FORCES

The extent to which the metropole could *control* the outcome and reception of Arctic explorations was limited, but Barrow exercised a significant amount of *influence* on the nature and direction of British Arctic science in the first half of the nineteenth century. Expeditions in the British North American Arctic were heavily shaped by the preferences of another key person: George Simpson (1786/1787–1860). In 1821 the HBC and NWC merged. The merger, which would be due for renewal after twenty-one years, ended a long-standing competition between the two companies, as the HBC gained a monopoly over the fur trade in all of British North America except for the St. Lawrence and the lower Great Lakes area.[24] At the time of the merger, the HBC was deeply indebted to the Bank of England. The new monopoly was unprecedented and offered an exceptional opportunity to create large profits on the fur trade. George Simpson was chosen as the governor in chief for the new HBC, and throughout his career he sought to shape its trajectory. Scientific activity and the search for a Northwest Passage were used by the HBC as a way to gain the power, respectability, and financial standing that Simpson desired.[25] But the HBC's merger and its new organizational structure were not easy to navigate, even if the company had not been struggling with its large debt. For the HBC, as well as the Danish crown and the KGH, the connection between knowledge of the Arctic and economic and imperial concerns are directly evident, not least in John Franklin's two overland expeditions.

Franklin was part of Buchan's 1818 expedition in search of the North Pole, but that expedition had had very little contact with the coast. Franklin's expedition between 1819 and 1822 was his first experience leading one overland. This was the first of three Arctic expeditions led by Franklin, referred to either as Franklin's first (Arctic Land) expedition, or the Coppermine expedition. This was where Franklin acquired the nickname "the man who ate his boots"—of the twenty-two members of the party, eleven died of starvation, murder, and, perhaps, murder for the purpose of cannibalism. The expedition was small, and Franklin left England with five men[26] and plans for engaging more in Canada. Their ship, the *Prince of Wales*, belonged to the HBC. The main objective of the expedition was geographical, aimed at "determining the latitudes and longitudes of the Northern Coast of North America, and the trending of that Coast from the Mouth of the Copper-Mine River to the eastern extremity of the Continent," in the hope of discovering a route to the Pacific. They succeeded in identifying the location of the Coppermine River and explored the coast eastward to adjust the older maps. In addition to the Northwest Passage, the expedition's goal was generally amending the geographical knowledge of the area and correcting the older maps with new measurements on latitude and longitude. This required them to travel into areas that were either unmapped or poorly mapped by Europeans and

Euro-Americans. As was often the case with expeditions that discovered areas already known to Indigenous peoples, Franklin's unmapped land was home to many, including the Denesuline (Chipewyan), Tlicho (Dogrib), and T'atsaot'ine (Yellowknives) peoples. In addition, there were the two trading companies, the HBC and NWC.

The British government had assured both the HBC and NWC that Franklin's expedition in no way intended to interfere in the dispute between the two companies, and the fur traders who assisted Franklin were promised reimbursement for the supplies or assistance they provided. While the companies had agreed to cooperate, in reality their bitter conflict caused problems for the expedition. In his published account, *Narrative of a Journey to the Shores of the Polar Sea*, Franklin acknowledged the assistance of the NWC and HBC in the introduction and throughout the narrative as individual figures associated with the companies aided them. Yet his frustration with the companies and the difficulties caused by their conflict was evident throughout. The failure of the HBC to provide the expected support was described as "unpleasant information," the "indecisive conduct" of the men referred to him by the HBC "was extremely annoying," and the HBC employees had been "backward in offering their services." At Fort Chipewyan, Franklin and his crew discovered that the bags of pemmican sent from the NWC had molded, and that the HBC had sent nothing as "the voyagers belonging to that Company, being destitute of provision, had eaten what was intended for us." They also had difficulties in getting a sufficient number of dogs from the NWC to carry the provisions required for their journey.[27] The two companies were unable (or unwilling) to assist the expedition as planned. These few examples serve to illustrate the difficulties Franklin's expedition had in navigating the wider sociopolitical context for their venture. The extent to which explorers could predict what to expect was limited, and this extended far beyond geographical and environmental uncertainties.

A key way Franklin's overland expedition, and others like it, handled this radical uncertainty was by relying on the labor and knowledge of Indigenous peoples. In Canada Franklin secured the assistance of the T'atsaot'ine people, in particular Chief Akaitcho (ca. 1786–1838). The T'atsaot'ine helped Franklin in countless ways: helping establish Fort Enterprise and giving Franklin's party advice on how to travel as safely and efficiently as possible. They also saved the surviving members of the expedition at Fort Enterprise in 1820–1821. Franklin was very aware of the fact that their survival had depended on Akaitcho's willingness to help them, often in situations when the trading companies had failed to

FIGURE 1.4. Portrait of Akaitcho and his son. The portrait was published in full color in John Franklin's narrative. *Source*: John Franklin, *Narrative of a Journey to the Shores of the Polar Sea*, 1823, 203. Courtesy of the Scott Polar Research Institute.

deliver on their promises of support. Franklin was also given advice, geographical and ethnographical, from the interpreter Augustus (d. 1834)[28]: "Upon the map being spread before Augustus, he soon comprehended it, and recognised Chesterfield Inlet to be the 'opening into which salt water enters at spring tides, and which receives a river at its upper end.' He termed it Kannæuck Kleenæuck. He had never been farther north himself than Marble Island . . . He says, however, that Esquimaux of three different tribes have traded with his countrymen, and that they described themselves as having come across land from a northern sea."[29] Augustus advised Franklin on the exactness of his map and highlighted what was not on it. For the areas where he himself had not been, Augustus pieced together a conjectural map based on information from the tribes he had traded with. It was a combination of ethnography and geographical navigation, which the historian of science Michael Bravo has argued can be understood as the "ethnographic process of exchange, performance, and translation because the surrender of the ethnogeographical knowledge (more than other ethnosciences) draws attention to its own contour as the source of geographical knowledge."[30] Franklin's map and later encounters with Indigenous peoples were in this way shaped by Augustus's knowledge of other tribes as well as the land.

Another example of the role of Indigenous peoples in the process of navigating Arctic was recorded in Franklin's narrative for August 25. Franklin wanted to continue the journey down the Coppermine River, but Akaitcho considered this much too dangerous and strongly advised against it. Franklin was upset, and when pushed on the issue, Akaitcho reportedly stated that "at the former place he had been unacquainted with our slow mode of traveling."[31] There were significant differences in the style of exploration between the Royal Navy, Indigenous peoples, fur traders, and voyageurs. The Royal Navy was used to expeditions with plenty of supplies, large boats full of food resources. Franklin's, however, had to travel light and carry their supplies on their backs and in their canoes. The total number of people in Franklin's party varied, but they were never a small party, often around twenty-seven. Judging from how the discussion was represented in his narrative, Franklin did not take Akaitcho's concerns seriously. At least, he was not happy to follow it:

> Akaitcho appeared to feel hurt, that we should continue to press the matter further, and answered with some warmth: "Well, I have said every thing I can urge, to dissuade you from going on this service, on which, it seems, you wish to sacrifice your own lives, as well as the Indians who might attend you: however, if after all I have said, you are determined to go, some of my young men shall join the party, because it shall not be said that we permitted you to die alone after having brought you hither; but from the moment they embark in the canoes, I and my relatives shall lament them as dead."[32]

Akaitcho eventually persuaded Franklin to postpone going down the Coppermine River. Instead Franklin sent expedition members George Back (1796–1878) and Robert Hood (1797–1821) in a light canoe to quickly survey the distance and size of the river. As Bravo argues, when explorers made use of Indigenous informants, they "allowed for their navigation routines to be interrupted by these encounters."[33] This exchange shows the reciprocal, although unequal, nature of the contact zone. Franklin relied on the assistance of Akaitcho and his people, and he knew he could not afford to alienate them. For example, he noted how he was "mortified to learn" that Akaitcho "had received some further unpleasant reports concerning us from Fort Providence, and that his faith in our good intentions was somewhat shaken." It was important for Franklin to retain Akaitcho's support. He later believed that Nicholas Weeks, a clerk at the North-West Company at Fort Providence, had spread these and other rumors about his crew, as the clerk supposedly believed the explorers were working against the interests of the trading company. Again Akaitcho worked to aid Franklin, as he in the presence of Weeks described how Franklin and his crew always had told the T'atsaot'ine people to "consider the traders in the same light" as themselves.[34] Akaitcho generally appears to have had very little trust in Franklin's abilities to control any situation. When he refused to accompany Franklin on his proposed journey down the Coppermine River and insisted that leaving so late in the year was effectively a suicide mission, he was helping Franklin to understand the geographical landscape. As the example with Weeks illustrates, he also helped Franklin navigate the complex sociopolitical landscape.

Akaitcho's interactions with Franklin are recorded through the writings of Franklin, Back, and other European and Euro-American peoples. It is important to remember that the translocation of Akaitcho's actions and intentions into Franklin's narrative was mediated in this way. Akaitcho was also aware of this. "'I know,' he said, 'you write down every occurrence in your books; but probably you have only noticed the bad things we have said and done, and have omitted to mention the good.'" Akaitcho knew that foreign explorers narrated their experiences following the return of their expedition, and that these representations could have significant implications for future relationships. According to Franklin, Akaitcho asked him to instead "represent the character of his nation in a favourable light to our countrymen."[35] Descriptions of Indigenous peoples made by explorers were used as evidentiary sources for anthropologists, and they could further be utilized when colonial administrators made decisions that impacted their livelihoods. Akaitcho hoped that Franklin's expedition could change the perceptions of the T'atsaot'ine people, and it is likely that Akaitcho's intended audience was the trading company administrators who he knew exercised a high level of control over the safety of and possibilities afforded to his people.

As seen through Franklin's two narratives, Akaitcho further revealed the extent to which Franklin relied on Indigenous knowledge to control the uncertainties associated with Arctic overland expeditions. As Back noted in Franklin's first narrative, "Mr Wentzel had taken away the trunks and papers, but had left no note to guide us to the Indians. This was to us the most grievous disappointment: without the assistance of the Indians, bereft of every resource, we felt ourselves reduced to the most miserable state."[36] During Franklin's first expedition, Akaitcho's people and Willard Ferdinand Wentzel (ca. 1780–1832), an experienced trader from the NWC, accompanied his party until they reached the Arctic Ocean, where they turned back. The party continued on with the two Inuuk interpreters, Augustus and Junius,[37] and a group of voyageurs. Perhaps because Franklin was now without Akaitcho's advice, the expedition continued beyond where was wise. William Williams (d. 1837), the resident governor of the HBC, had promised to forward provisions in the spring to Fort Enterprise. But the company turned out to be of little help. As the historian Anthony Brandt has written, "The HBC was a business, and it was all business."[38] The conflict between the HBC and NWC was at a high point, and there were few extra resources to assist an expedition such as Franklin's. When it became clear that no food sources were available, the party divided for the first time, with Back and a small group heading out toward Fort Enterprise to bring the supplies they thought would be there. But they found no food. Back continued from the fort to find Akaitcho. With no sight of Back, the party divided again. John Richardson (1787–1865), the surgeon to the expedition, believed Michel Teroahauté (d. 1821), an Iroquois voyageur, had murdered a voyageur for cannibalism, and when Hood was found shot dead, Richardson shot Teroahauté. Between murder and cannibalism, Richardson and the seaman John Hepburn (1794–1864) were the only survivors of this division of the expedition. More people died at Fort Enterprise before they were rescued by the T'atsaot'ine. The survivors at Fort Enterprise gradually improved "under the kindness and attention of our Indians." On November 26 they arrived "in safety at the abode of our chief and companion, Akaitcho. We were received by the party assembled in the leader's tent, with looks of compassion, and profound silence, which lasted about a quarter of an hour, and by which they meant to express their condolences for our sufferings. . . . The Chief, Akaitcho, shewed us the most friendly hospitality."[39] While it should not be forgotten that Franklin's two narratives certainly included many instances of racist and paternalistic language, his *Narrative of a Journey* is distinct in the extent to which it credited the labor and knowledge of Indigenous peoples, in particular Akaitcho, and to which Franklin emphasized the hospitality and generosity of the T'atsaot'ine people.

Akaitcho appears again in Franklin's narrative from his second expedition between 1825 and 1827, but during this expedition, he was unable to assist them as he had before. War had broken out between the T'atsaot'ine and the Tlicho peoples in 1823, and when peace was reestablished Akaitcho did not want to risk restarting the conflict by returning to where the war had taken place.[40] Franklin expressed his deepest sympathy for the situation: "Such sentiments would do honour to any state of civilization, and show that the most refined feelings may animate the most untutored people. Happily we were now so circumstanced as to be able to reward the friendship of these good men by allotting them from our stores a liberal present to the principal persons."[41] Most of the men that had saved Franklin and his party had sadly been killed. The T'atsaot'ine were a small band, and thirty-four of them had been killed in the war with the Tlicho, amounting to about one-fifth of the entire band. Moreover, because of the merger between the HBC and NWC, Fort Providence, where the T'atsaot'ine had been trading, closed in 1823, and they now had to trade out of Fort Resolution, which was already being used by the Denesuline people.[42] Akaitcho becomes a central figure again during Back's overland expedition in search of John Ross's party. In 1836 George Simpson and the HBC issued a statement that future expeditions would prioritize the use of Indigenous methods to ensure successful travels in the Arctic.[43] But we can see hints that individual explorers had earlier been aware of the benefits of adopting Indigenous methods and establishing friendly relationships with Indigenous groups during their expeditions. Indigenous knowledge and methods were a central part of shaping the overland expeditions, perceptions of the land, and knowledge of Indigenous peoples.

Franklin's first overland expedition had many difficulties fulfilling its broad and ambitious scientific aims. They had been forced to leave their collected specimens that had not already been sent back with the HBC due to hunger and death on their return journey. Aside from the geographical objective of the expedition, the scientific areas of interest involved making general observations related to natural history, most significantly magnetism, including the phenomenon of the aurora borealis, "observations that might be likely to tend to the further development of its cause, and the laws by which it is governed."[44] Franklin's narrative included several appendixes of scientific data, in addition to what was detailed in the body of the text. Richardson carried out detailed meteorological, geological, zoological, and botanical observations. Hood had, before his passing, made several magnetic and meteorological observations. *Narrative of a Journey* included an appendix "Notices of the Fishes," authored by Richardson, which included beautifully colored illustrations of fish. Richardson stated that his reasons for including this much detail on fish was to advance the science of ichthyology. Rather than deciding himself if a fish had been described before, he included details of all the fish he observed. He claimed that with the exception of "one or two instances" all the descriptions were written on the spot, seeking to emphasize that his representation of the Arctic animals was free of any type of observational or methodological bias. This was particularly important because they had lost many of their collected specimens.

The zoological appendix was organized by Joseph Sabine (1770–1837), based primarily on Richardson's observations and making use of George Cuvier's (1769–1832) taxonomy. Cuvier expanded the Linnaean taxonomy through a comparative anatomy that emphasized the internal structure and function of animals. Sabine highlighted the heroic nature of Richardson's Arctic science, writing "Neither privations, fatigue, not the inclemency of the Arctic winters retarded his exertions, which have been particularly marked by the extent of the collections of specimens which have been received from, or brought home by, him." In spite of Sabine's flattering words to Richardson, there was a strained relationship between the two, and Richardson's actual collections were more extensive than what appeared in Sabine's narrative.[45] Richardson published his additional observations, made during both the first and second Franklin expeditions, as *Fauna Boreali-Americana, or, The Zoology of the Northern Parts of British America*, with the assistance of the illustrator and ornithologist William Swainson (1789–1855) and Reverend William Kirby (1759–1850). It was published by John Murray "under the authority of the right honourable the secretary of state for colonial affairs." Swainson completed the drawings of the birds, in addition to organizing the ornithological section. The section on insects was done by Kirby, and a professor of botany at Glasgow, William Hooker (1785–1865), organized the botanical section. Hooker and Thomas Drummond (1780–1835) published additional zoological and botanical material from the expedition. The cost of the twenty-eight illustrations in Richardson's book, in addition to illustrations not published, had been supported with £1,000 from the British government.[46] Unfortunately, the majority of the illustrations in Franklin's 1823 narrative and all of those in Richardson's book were monochrome.

In the appendix to Franklin's narrative, Sabine expressed his surprise that so little was known about the fur-bearing animals of the region, considering their value in the fur trade. The economic benefits of exploring and cataloging the animal and plant life in British North America were not lost on Sabine. Richardson himself noted the incompleteness of the observations made during the expedition and hoped future observers could correct and add to his work. The lack of structure, or a set methodology, had its downsides, yet the differences between Franklin's first and second expeditions were significant. The second between 1825 and 1827 was much better prepared for the undertaking. As Richardson described, "Previous to our setting out on the Second Expedition, Sir John Franklin addressed letters to many of the resident chief factors and traders of the Hudson's Bay Company, requesting their co-operation with our endeavours to procure specimens of Natural History, and their ready acquiescence with his desire was productive of much advantage to us."[47] Because Franklin was much more aware of what to expect, his second expedition went better than his first. This coincided with Frederick William Beechey's expedition to explore the Bering Strait in the HMS *Blossom*. The two parties were supposed to meet, but Franklin had learned from the catastrophes of his previous expedition and turned back in order to ensure the lives of his crew. In particular, the second expedition was made possible by cooperation with the newly amalgamated HBC. During the first, the HBC and NWC were too preoccupied with their feud to fully assist the expedition, but from the mid-1820s on the HBC was interested in contributing to Arctic explorations. The Northwest Passage was in many ways a prestige project for the company. Support for exploration and scientific investigations was used by the company to improve its image by showing that the it was engaged in other than primarily economic pursuits—a theme which will be further elaborated on in the next chapter. Both the HBC and the British government were concerned about the geopolitical situation of the Arctic. The company was also interested in the potential financial benefits from cataloging the fur-bearing animals, and, of course, ensuring its new monopoly on trade was upheld.

Taken on its own, Franklin's first expedition was largely a failure. But its hard-won lessons were passed on to future groups of explorers, by setting important precedents for the future. Significantly, it gave the British Royal Navy much needed experience in undertaking overland expeditions in the Arctic, although they were not convinced that this small type of expedition was preferable to the large-scale two-vessel ones. But as was frequently the case, these lessons were gained through the labor of peoples who did not share in the possible long-term benefits. Even in the short term there were delays in compensating the T'atsaot'ine for the support they had provided Franklin's expedition. Franklin was disappointed to learn that when the NWC and HBC merged, "the goods intended as rewards to Akaitcho and his band, which we had demanded in the spring from the North-West Company, were not sent." According to Franklin, however, Akaitcho was not surprised and did not blame Franklin: "'The world goes badly,' he said, 'all are poor, you are poor, the traders appear to be poor.'"[48] It is interesting to note that Franklin kept the word *appear* in Akaitcho's quotation, as it further insinuated that the trading companies had been falsely representing themselves. Taken together, Franklin's *Narrative of a Journey* is an example of how overland expeditions managed geographical uncertainty. It also highlights the international nature and collective work of exploratory missions, as the British explorer relied on several groups of fur traders and Indigenous peoples. In particular, Franklin's narratives reveal in many different ways the extent to which Indigenous peoples such as Akaitcho and Augustus shaped the direction and outcome of expeditions, and the way explorers relied on Indigenous peoples to help control the uncertainties associated with Arctic exploration.

DENMARK IN GREENLAND: WILHELM AUGUST GRAAH

If looking to the first British expeditions following the Napoleonic Wars reveals the efforts to discover what could be accomplished with large-scale ventures to the Arctic in the first part of the nineteenth century, then these expeditions and those of the Danes to Greenland tell a different story. There was a marked difference in the availability of Danish and British imperial funds in this period. These financial resources shaped the organization of Arctic explorations, both in terms of the voyage structure and with regard to how many expeditions it was possible to send out. The first postwar British expeditions were shaped by uncertainty, both in terms of what to expect when traveling through the icy north and what could be accomplished in this new natural laboratory. The first Danish expeditions to the Arctic after the Napoleonic Wars differed in three significant ways. First, they were not focused on finding a Northwest Passage, but sought instead to find traces of the lost Nordic colony, while simultaneously ascertaining what resources could be extracted for trading purposes in the Danish Empire. Second, the expeditions were small and limited to a handful of people. Third, the involvement of the KGH—as with the expeditions organized by or in conjunction with the HBC—changed the level of uncertainty both with regard to what could be accomplished by the expeditions and what to expect from the environment.

Between 1828 and 1829 the Danish explorer William August Graah led an expedition to the east coast of Greenland. Graah's mission was geographical, but he was searching for something very different than a trading route. Aside from surveying and mapping, a powerful strategy for asserting imperial dominance was establishing a historical link to a region. In the case of Greenland, the Danish crown was keen to establish a record of continuous presence, centered around the mythological lost Nordic tribe. The existence of the lost tribe was a way to justify the Danish imperial presence in Greenland by giving evidence of a long-standing historical settlement, and Graah's expedition should be seen within the context of these imperial efforts. When Napoleon I of France in 1806 launched the Continental System (*Blocus continental*) as a way to isolate Britain from the rest of Europe in response to the British naval blockade of the French coasts in May 16, 1806, Britain's first response was an attack on Denmark. Although Denmark-Norway had attempted to stay neutral, there was significant pressure from Napoleon to pledge the Danish Navy for his use against Britain. Britain initiated an attack on Denmark in July 1807 with the purpose of claiming control over the Danish fleet. The British Navy bombarded Copenhagen between September 2 and 5, during which over a thousand buildings burned. Denmark surrendered on September 7, and Britain took charge of its navy. In 1813 Denmark went bankrupt, and in 1814 was forced by the Treaty of Kiel to pass governorship of Norway to the Swedish crown and give up Helgoland to Britain. Norway disputed this and declared their own sovereignty at a national assembly at Eidsvoll on May 17, 1814. After the wars, in contrast with Britain, the Danish crown had no funds or resources for expensive exploration ventures. But the Treaty of Kiel formally gave Denmark ownership of three former Norwegian or Danish-Norwegian areas: the Faroe islands, Iceland, and Greenland—something the now-sovereign Norway unsuccessfully contested for years.[49]

Graah's expedition specifically aimed to find traces of the so-called East Bygd (*de Gamles Østerbygd*) while making scientific observations, collecting natural history specimens, and establishing friendly links with the Indigenous groups living along the coast. Upon completion of the expedition, Graah published an account of his venture, *Undersøgelses Reise til Østkysten af Grønland* (1832). The narrative contained five appendixes: the true site of the East Bygd, and zoological, botanical, meteorological, and other scientific observations.[50] The four latter categories were similar to those of the British expeditions, but the first category shows the variation in focus of Graah's expedition. At the time it was thought that the lost Nordic tribe had settled in the East Bygd, but the exact location was uncertain. The existence of the lost Nordic tribe was a way to justify the Danish imperial presence in Greenland by giving evidence of a long-standing historical settlement, but it was also a topic of interest to Arctic researchers, especially those occupied with historicism and human development.[51] For example, the British explorer Scoresby had been keen on finding traces of the lost Nordic tribes when he had surveyed East Greenland in 1822. In his narrative, *Journal of a Voyage to the Northern Whale-Fishery: Including Researches and Discoveries on the Eastern Coast of West Greenland, Made in the Summer of 1822, in the Ship Baffin of Liverpool* (1828), Scoresby noted that the question of the lost Nordic tribes was probably the most interesting part of his journey for the broader public.[52] Scoresby did not actually meet any other humans during his voyage, but he drew his conclusions primarily from his observations of the remains of abandoned dwellings. These indicated, he believed, that their former inhabitants were not only Esquimaux but linked to the lost tribe.

Scoresby claimed that he was the first British subject to undertake a survey like this in the region. Graah's narrative referenced Scoresby, together with a lengthy discussion of other voyages to the region. However, Graah pointed out that

> although he [Scoresby] did succeed in landing at several points of the East coast, he did so at a much higher latitude than where the ancient colonies were to be looked for, and at points where, it is probable a landing might in most years be effected. In fact, long before his time, the portion of the East coast between 70° and 75° latitude had been visited by Danish, Dutch, and English whalers. His merit consists in having furnished an interesting account of this part of the coast, and a more authentic chart of it than any we before possessed, though he, unquestionably, is in error, where he asserts that, by keeping close in shore, one may sail along the whole East coast from lat. 70° to Cape Farewell. Danel's, Olsen's, and Egede and Rothe's expeditions, prove the fallacy of this opinion.[53]

Graah drew on a comparative study of past narrative accounts to evaluate Scoresby's claims. This type of retrospective evaluation of findings from past expeditions formed a key part of the evidentiary material of explorers, in the Arctic and elsewhere. This points to a broader question of what counted as evidentiary material and the labor of individual explorers to portray themselves

Figure 1.5. An illustration of sailors in a umiak, a type of boat used by William August Graah on his expedition, which was printed in full color in his narrative. *Source*: Wilhelm August Graah, *Undersøgelses-Reise til Østkysten*, 1832, plate 1. Courtesy of the Scott Polar Research Institute.

as knowledgeable about all parts of the Arctic. What counted as evidence was broadly defined, including the remains of buildings or artifacts as well as cultural practices. As such, *Undersøgelses Reise* included detailed ethnographic observations that drew upon both historical and contemporary evidence. Comparing your own observations to that of past explorers and using their work to corroborate your findings was an effective way to show the progress made by your work. Conversely it was also a way to reject or circumvent reports that did not align with your findings. It is useful to think about this approach as a type of ethnography, or historical geography, that interwove archival evidence with firsthand observations in a conversational format. In this instance Graah concluded that the East Bygd was not on the east coast of Greenland, as "has been asserted by a no less scientific navigator than Scoresby," but on the west coast.[54] Graah's expedition found no trace of the lost colony but determined that the East Bygd was only called "East" in its reference to another *bygd*, both on the west coast. Past explorers, Graah concluded, had been searching in the wrong spot.

Graah's expedition left Copenhagen in the ship *Hvalfisken* on Sunday March 30, 1828. *Hvalfisken* was a brig, a fast and easily maneuverable ship, which belonged to the Greenland Board of Trade. Upon arrival in Cape Farewell they changed to an umiak, a type of boat used by Indigenous Greenlanders, also called a "wife boat" (konebåd). The umiak was well suited for long-distance travel such as this, as it was small and easy to maneuver—if you knew how to use it. Graah's expedition was small and relied heavily on the cooperation and active support of Indigenous Greenlanders. The core team from Denmark consisted of two naturalists, the geologist Christian Pingel (1793–1852) and the botanist Jens Vahl (1796–1854), the superintendent of what was known as the Colony of Frederick's Hope, Jens Mathias Matthiesen (1800–1860), an unnamed sailor who was also the cook, together with several Greenlanders commissioned at different points during the expedition. The main guide to the expedition, Ernenek,[55] was hired upon arrival in Greenland. The importance Graah placed on finding the right guide is evident in the lengths that Graah went to in order secure Ernenek's help. While still deliberating the offer, Ernenek asked Graah to give him time to speak to his family and to meet him elsewhere later. Graah wanted to make sure that Ernenek would agree to his offer, so he decided to travel with him to Nennortalik to secure his family's approval.[56] Ernenek's expertise was necessary for the success of the expedition, and Graah knew this. After meeting with Graah, Ernenek's two wives agreed to accompany the expedition.

The official instructions requested that Graah use his leisure time between his arrival in Greenland and his departure up the east coast to chart the coastal district of Qaqortoq (Juliana's Hope). Graah reached farther north on the eastern coast of Greenland than any other European had before, and the map of the coastline he produced was one of the most significant results from the expedition. Graah also collected animals and minerals for the Danish Royal Museum, and plants and seeds for the Botanic Garden in Copenhagen. In instances where it was not possible to collect a specimen, the official instructions requested that they produce colored drawings "with a view to their insertion in the Flora Danica."[57] Pingel had no part in the official instructions but accompanied the party to Greenland on a somewhat independent basis with the purpose of carrying out geological research. As the expedition botanist, Vahl was instructed to collect materials for Flora Danica by the professor of botany at Copenhagen University Jens Wilken Hornemann (1770–1841). The production and publication of Flora Danica was one of the key projects for cataloging the natural history of the Danish Empire in the eighteenth and nineteenth centuries. Its publication was an ongoing project between 1761 and 1883 and contained a total of 3,240 copper plates in fifty-one volumes and three appendixes.[58]

In the eighteenth century, Carl Linnaeus (1707–1778) developed a systematized method for cataloging the natural world, known as binomial nomenclature. Although Linnaean taxonomy was practical and fairly straightforward, the varying amounts of information provided, together with the specimens or illustrations, meant that some plants had been mis-categorized, leaving *Flora Danica* in a disorganized state. Jens Vahl's father, Martin Vahl (1749–1804), had been a professor of botany at Copenhagen University and coeditor of *Flora Danica*. A student of Linnaeus, he wanted to bring order to the collection of data and establish a catalog of all the plants in the world organized after Linnaean principles. The Linnaean system was widely adopted by naturalists in their efforts to classify the entire globe. For example, the German naturalist Alexander von Humboldt (1769–1859) combined this system with his own ideas about how the scientific traveler could systematically catalog the natural world. Humboldt proposed a *physique du monde*, a universal natural science of the Earth based on systematized observation, measurement, and experiment. As Martin Rudwick has written in *The Meaning of Fossils* (1976), "'Natural History' was still, as it had been for Linnaeus and Buffon in the eighteenth century, the systematic ordering of the whole range of diverse natural entities."[59] Arctic explorers, as with those in other areas of the world, were faced with the problem of categorizing a natural world that sometimes looked vastly different from what you would find in the imperial metropole. Graah and his two naturalists worked within the framework of Linnaean taxonomy when they collected specimens during their expedition to Greenland. Upon the expedition's return to Denmark, Hornemann classified, again in the Linnaean system, the specimens Graah had brought home with him.

Jens Vahl's collection from the voyage added a substantial amount of new knowledge to *Flora Danica*. Graah described how the inhabitants of the village Nennortalik gave Vahl a nickname that described his preoccupation with collecting natural history specimens: "They gave thus to Mr. Vahl the name of 'Piniartorsoak,' i. e., the diligent earner, not because be exhibited any great skill at catching seals, (which the word literally signifies,) but because they observed him to be constantly in chase of gnats and flies, intended to be added to his collection of insects."[60] Vahl did the primary part of collecting plants, birds, and insects, as well as making meteorological observations, whereas Graah focused on making observations on magnetism and the aurora borealis. The atmospheric phenomenon of the northern lights was a key area of interest. What was its cause variations, its movements and shapes, and how was it related to other meteorological phenomena, magnetism, and sound, were key questions for researchers in this period. Many of Graah's descriptions highlighted the beauty of the Arctic. For example, his account of the northern lights showcased its splendor. He wrote that they were "a remarkable and beautiful phenomenon of which the inhabitants of the greater part of Europe can form no adequate conception" and described two types of light in details.[61] The first type could be found uniformly between magnetic ESE and SW, or WSW. He estimated that the highest point of the light's arc in the magnetic south was between 10° and 20° above the horizon, from where rays of light spread out. This type, he believed, usually preceded a great temperature change, while he linked the second type to barometrical changes. He described this type in vivid terms, emphasizing the wonders of the phenomenon: "The other sort of northern light, which, still more than the former, seems to stand in connexion with barometrical changes, flits from place to place in the semblance either of light luminous clouds agitated by the wind, and through which the light appears to diffuse itself with a sort of undulating motion, or of flaming rays, flashing, like rockets, across the firmament, most commonly upwards in the direction of the zenith, or, finally, like a serpentine, or zig-zag belt of vivid, undulating light, frequently coloured, which at one moment is extinguished, and the next relit."[62] Graah described a version of this second type, which he considered to be most beautiful, namely the corona, described as "a luminous ring near the zenith, of from 2° to 3° in diameter, with rays diverging in every direction, like prolonged radii, from its centre." The corona lasted only a few seconds at a time, appearing like an explosion of light on the sky. According to Graah's observation, it primarily appeared "to the east of the meridian, at an elevation of from $81\frac{1}{2}°$ to $82\frac{1}{2}°$ above the horizon."[63] The position of the corona was determined first by reference to the stars around its center. Graah used the horary angle, or hour-angle—a way of determining the altitude of objects in the sky—which gave the distance from the meridian in time to determine the azimuth and altitude of the center of the corona. In addition to determining the positions of the aurora, Graah also addressed the question of its sound. He believed that the "low, hissing noise" that sometimes accompany the northern lights was due to a combination of movements in the ice and the wind moving over the snowy landscape. To examine the effect of northern lights on magnetism, he suspended a magnet from a silk fiber during their more vivid occurrences and found no effect. Conveying the color of the Arctic was important for scientific purposes, but it also factored in how a reader of the travel narratives would imagine the region. Graah's Arctic landscape was one of color; atmospheric phenomena, vegetation, animals, glaciers, and ice, as well as Inuit were all described in vivid language. The illustrations in his narrative, prepared from his sketches and finished back in Denmark, added to this colorful portrayal of Greenland.

Here it is instructive to point out two key functions of the visual material produced during voyages of exploration. First and foremost, visuality formed an important part of the imperial project. Scholars such as Daniella Bleichmar and Sujit Sivasundaram have emphasized the significance of visuality in conceptualizations of the imperial periphery. Such an approach draws our attention to the broader context of the visual language of narratives and the production and function of visuality. As Sivasundaram noted with reference to the visual politics of high empire on the island of Sri Lanka, "Colonial photographs harmonize with the metaphors and imaginaries of colonial intellectual disciplines and the markets and finance that sustained empire."[64] The imperial metropole and periphery, race, gender, and class were created by their relation to each other in complicated interactions evident in both text and image. The copperplate engraved illustrations included in Graah's narrative reveal the composite nature of what counted as evidentiary material: there were visual representations of people in situ handling objects and performing household tasks (ethnography), ruins (archaeology), and flora and fauna (natural history). In addition, the visual language of these illustrations spoke to Graah's direct firsthand observation of the Arctic.

The publication of Graah's colored illustrations was made possible by a financial subsidy from the government, as well as developments in printing technologies. While Ross's narrative from his 1818 expedition included several hand-colored images (including one of the Croker Mountains), Ross's book was expensive, and its images were generally not reproduced in the long summaries circulating in the press. The visual representations of the Arctic, in this way, were shaped by complex financial and social factors, in addition to developments in the printing press, and these differed between the national contexts. Graah's illustrations are suggestive of multiple functions of travel narratives: as scientific literature, as commodities for consumption in the metropole, and as an account of the Danish imperial presence in Greenland. Images were a central way of connecting the metropole with the empire. Visuality traveled over oceans, to make the world knowable to both the scientific and a broader audience, and were an important part of imperial projects.

Graah's narrative was translated into English and published in 1837. Its publication was delayed by the sudden death of the translator, George Gordon

Macdougall, which in turn gave James Clark Ross an opportunity to add footnotes to the book. The editorial noted that they had decided to keep Graah's "homely" style of writing rather than changing it to "the more usual forms of expressions."[65] This reflects an interesting "othering" of Danish explorers in Britain. It is easy to generalize and create a dichotomy between European and Euro-American explorers on the one hand, and Indigenous peoples on the other. But there was no sense of unity among explorers across national boundaries, as the following chapters explore, and international collaborations were often fraught. Perhaps even more significantly, there was (and is) no singular Arctic Indigenous culture with which historians can neatly create a dichotomy. Such considerations are significant for many obvious social and political reasons, but also if we are to understand more clearly the function and nature of science in the Arctic. As Sivasundaram reflects, by carefully examining what went into establishing connectivity, we can point out not just what was made visible, but also what and who were erased in the process.[66]

The English translation was made for the Royal Geographical Society of London, and contained the appendix and Graah's original chart from the expedition. This version did not include the original colored illustrations. This points to a significant difference between the illustrations in Ross's and Franklin's narratives and Graah's. In the former, the illustrations of Indigenous peoples were highly stylized, more like portraits. Graah's images appear like vignettes of what daily life in Greenland was like, capturing for the reader what it was like seeing Inuit in situ going about their daily routine. This was mirrored in his style of writing. While Linnaean botany focused on producing images that were idealized composites rather than realistic depictions of unique plants, and this was the style of images produced for the *Flora Danica*, the illustrations of Indigenous Greenlanders in the narratives of both countries were meant to be a visual representation of actual living people. The illustrations of Indigenous Greenlanders and their living quarters were not in the style of what the historian of art Martin Kemp described as Goethe's "Ur"-form—similar to what the historians of science Lorraine Daston and Peter Galison have termed "truth-to-nature"—where "the 'leaf archetype' has no existence as such. Rather it is a supreme exemplar of the kind of organizing and generative template (or metaphysical 'form') which the discerning student can recognize as expressing the principles of unity on which God has constructed the manifold varieties of nature."[67] Instead the illustrations in Graah's narrative are more the style of what Bleichmar has described as travelers functioning as a surrogate eye for the reader. The colored illustrations of Graah's narrative and *Flora Danica* are suggestive of a different type of visual epistemology than the British. Graah's illustrations were linked to an ethos of knowledge dissemination also present in the *Flora Danica* and the later *Grundtvigian Folkehøjskoler* (folk high schools), which started in Rødding in 1844.[68] Copies of *Flora Danica* were given out to what can usually be considered epicenters of knowledge dissemination, such as parishes and libraries, available for perusal by people who could not afford a copy of their own, to acquaint all readers with the Danish botanical kingdom.

Graah recounted in detail the customs and morals of Inuit from the east coast of Greenland. His descriptions were very positive, praising their love for their children, sense of honor, and adversity to saying something that could offend, "things of which they, however, entertain notions widely different from ours." Graah was responding to the account of the inhabitants of the east coast who "from time immemorial, they have been cried down as infinitely more savage and cruel" than the inhabitants of the west coast. However, Graah did not appear to disagree with the extremely negative and racialized accounts made by writers such as Hans Egede regarding Greenlandic Inuit customs.[69] According to Graah, Inuit from east Greenland were honest, never raped or plundered, only stole if it was a matter of life or death, and were hospitable, forgiving, and forbearing. Yet in the same passage he claimed that "their worst faults are—ingratitude, a total want of sympathy for the distressed and destitute (those excepted who are related to themselves), and cruelty to dumb animals."[70] Despite his description of Indigenous Greenlandic customs and morals, his accounts of his engagement with them, and the value he placed on people such as Ernenek, Graah was still unable to let go of the negative and racialized language so prevalent in European accounts of Inuit during this period.

Also of interest to Graah were the types of clothing worn by east coast Inuit. He noted that "in Summer, when at home, or in Winter, when in their heated earth-huts, a scanty pair of breeches constitutes their entire dress."[71] The narrative included an image that shows a striking difference in the amount of dress worn between the two women and the man, bolstered by the text. The identity of the portrayed is unclear, but it was likely the three people Graah shared living quarters with during the winter of 1829 and 1830, a woman called Sorte (Black) Dorthe, an unnamed woman from Nennortalik, and her partner Ringeoat.[72] The two women were almost naked, and performing domestic tasks with their breasts clearly visible. By contrast, the man was portrayed fully clothed, sitting by a fire. While the choice to portray them naked could be a matter of attempted realism,

FIGURE 1.6. An illustration of living quarters in Greenland. The illustration was printed in full color in Graah's narrative. *Source*: Wilhelm August Graah, *Undersø-gelses-Reise til Østkysten*, 1832, plate 7. Courtesy of the Scott Polar Research Institute.

it is suggestive that Graah did not portray them in the other styles of clothing he described in detail in the narrative. Anne McClintock has proposed the concept of "European porno-tropics" within which "women figured as the epitome of sexual aberration and excess."[73] From this perspective, the choice to portray the women naked was linked to ideas of Indigenous peoples as more sexual and therefore more primitive. But Graah's narrative does not straightforwardly lend itself to this conclusion. Elsewhere Graah emphasized the fidelity within married couples in East Greenland: "When married, they lead, in general, a reputable life together." Further, he suggested, young unmarried women "have many self-denials to endure, in order to avoid . . . placing in jeopardy their reputation, or their life."[74]

There is a tension in Graah's writing between his portrayal of Inuit as civilized with a detailed and interesting culture and his emphasis on their perceived primitiveness, including sexual promiscuity. This was, as will become clear in later chapters, a significant and problematic feature of Danish imperialism in Greenland. Graah's perception of Greenlanders is especially important because upon his return to Denmark, he was appointed director of the KGH. The historian Klaus Georg Hansen has described the impact of Graah's leadership as a radical change in the Danish administration of Greenland aimed at improving the living conditions of Greenlanders: "Graah changed the character of Danish colonialism in Greenland through the 1830s from parasitic to intensive colonization, with the improvement of the Greenlanders as its new primary goal."[75] However, as the next chapter shows, what the Danish imperial authority in the Greenland considered "improvement" was a highly problematic and violent aspect of the civilization project, as was also the case with the HBC in Canada.

Graah's narrative was well received in Denmark, and a review was published in the journal *Maanedskrift for Literatur* with long summaries of each section of Graah's book. The review was written by Pingel, who was perhaps not fully impartial, given his, albeit informal, involvement in the expedition. Throughout the review he referred to Scoresby's expedition in Greenland, as well as that of Sabine. Knowing that the natural world of the kingdom was intimately linked with imperialism, Pingel congratulated the Danish government not only for supporting the expedition but also for supporting its publication of the results so that the purchase price could be low,[76] part of the ambition to catalog and gain knowledge about the empire available to the entire Danish kingdom (a project also reflected in *Flora Danica*).

The period after Graah's exploratory mission to the east coast of Greenland was a quiet one for Danish expeditions, with no major ones organized. However, the KGH was busy with trading, and the newspapers brought regular news of the influx of goods for sale at auction. The natural history of Greenland also continued to be classified. After accompanying Graah on his expedition, the botanist Vahl remained in Greenland until 1836, undertaking botanical research and contributing to *Flora Danica*, as well as to Hinrich Rink's *Naturhistoriske Bidrag til en Beskrivelse af Grønland* (1857). His observations were also recorded in the periodical *Det Kongelige Danske Videnskabernes Selskabs Naturvidenskabelige of Mathematiske Afhandliger.*[77] After completing his expedition, Graah joined KGH's committee between 1831 and 1850, where he played a key role in shaping its policies and practices. Graah's expedition established the location of the East Bygd and charted a long area of the coastline, heavily supported by Ernenek. Graah's view of the many other Indigenous peoples who supported the expedition is significant, especially because of his later career trajectory. His narrative shows the tension between how explorers relied upon the assistance of Inuit to travel and survive in the Arctic, and the negative stereotypes harbored against them.

UNCERTAINTY AND THE EARLY ARCTIC EXPEDITIONS

From Ross's expedition on, the experiments and observations made during Arctic explorations reflect a form of scientific and imperialist research, which was significant regardless of the employment of its results by the scientific researchers in the metropole. What they were not was a unified scientific practice, or "Arctic science." The early British Arctic explorations were shaped by an acute uncertainty, which, combined with the emphasis on broad collecting practices and imperial ambitions, created high expectations for what they could accomplish scientifically. The ethos was to catalog and collect as much as possible, typically following the Linnaean system of taxonomy, but results were not particularly systematized. As was the case with both Ross's and Graah's voyages, the process of making geographical knowledge during Franklin's expedition elucidates the multidirectional construct of the Arctic through the narratives. The Arctic, Indigenous peoples, scientific practices, and the character of the Arctic explorer were all constructed simultaneously through their narratives and the reception of these accounts. The case of Ross illustrates the detrimental effect narrative choices could have on explorers' careers. Ross made a mistake with the Croker Mountains, but this was not his main error. Because of the way he constructed his persona, with a plurality of hidden voices in the narrative, he became an easy target for Sabine's criticisms. Sabine delegitimized Ross's authority as an observer of Arctic phenomena, both on the basis of the Croker Mountains as well as his supposed plagiarism. As the next chapter shows, Ross did succeed in venturing to the Arctic again, but not as part of a Royal Navy expedition. His nephew and Sabine, on the other hand, continued to have active careers with the government.

The first expeditions to the Arctic following the Napoleonic Wars were shaped by uncertainty and a marked absence of standardized methods for scientific research in the Arctic, except for a shared emphasis on the collection of as much as possible related to the natural world. The official instructions to expeditions were attempts to control the results, and they could be used against the explorer later, as was the case with Ross. However, because of the character of Arctic explorations and the uncertainty associated with them, the metropole could not control the results. While Franklin's first expedition has been treated in detail by other scholars, it is still significant to consider, because—like Ross's expedition—it shaped subsequent expeditions and was continuously referred to throughout the nineteenth century. Because the KGH and the HBC had long-standing presences in the Arctic, the uncertainty about what to expect was less marked in expeditions co-organized with the trading companies, although, shown by Franklin's expeditions, it did not safeguard the explorers from danger and failure. Franklin's second overland expedition in search of the hypothetical Open Polar Sea went much better than his first. They were better prepared, and the HBC cooperated, partly because the expedition could help expand the its trading capacity, further its image, and block Russian expansion into the region.[78] Likewise, the KGH had a vested interest in maintaining Danish authority in Greenland. In the case of both Graah and Franklin, Indigenous peoples shaped the trajectories of the expeditions, and by extension, the results they produced. This further illustrates the discord between the desires of the metropole and the reality of life in the Arctic periphery, which in turn challenges the metropole–periphery divide. The metropole, in fact, could not determine the results of the Arctic explorations.

Science in the Arctic was shaped not only by the training and abilities of the explorers, but also by interactions with Indigenous peoples, the financial context of the expeditions, and the unpredictability of the environment. A unifying feature of all the expeditions examined in this chapter is the disunity of scientific practices, as well as the close links between the textual and visual representations in the narratives and the perceptions of the Arctic, scientific practices, and the character of the explorer. Scientific practices in the Arctic were not simply transferred in a diffusion model from the elite communities in the metropole, but rather negotiated as their own genre.

CHAPTER 2

Financial Opportunities in the Arctic

The failure of a fourth attempt within these seven years, at the discovery of a North West passage, raises the very interesting question, how long such a course of unpropitious adventure is to be persisted in, and how often the appalling risk of brave men's lives is to be repeated?

— Anonymous, "The Failure of a Fourth Attempt within These Seven Years, at the Discovery of a North West Passage," 1825

Following William Parry's third unsuccessful expedition in search of a Northwest Passage between 1824 and 1825, several commentators questioned the logic of continuing this pursuit. The sentiment expressed in the editorial above was not unique. Almost a decade of Arctic expeditions seemed to have generated more questions than answers. Did a polar ocean, or the Open Polar Sea, actually exist, and how could it be reached? Was there a passageway through the frozen waters around Baffin Bay? If a route were found, would it even be economically advantageous to use it? While these concerns were not new, they were now being raised with increasing intensity. As was further noted in the editorial, "It therefore, we say, becomes a serious duty for the King's Government to weigh well the reasonings for and against another of these perilous and expensive trials."[1] However, governments were not the only possible sponsors of Arctic explorations.

As earlier in the century, there was a steady interest in Arctic explorations in the late 1820s and 1830s, but visions of what could be accomplished by exploring the region were changing. Whereas the dream of a fast sea route to the Pacific through the archipelago had played a key factor in the British Royal Navy's eagerness to send out expeditions after the end of the Napoleonic Wars, it was increasingly becoming evident that the financial gains inherent in finding the Northwest Passage were just that, a dream. Even if the complete passage could be traced, the cost of using such a route, both financially and in human lives, appeared to outweigh its commercial benefits. With each failed expedition, the question of whether additional ones could be justified became more pressing. In the British context, the disillusionment with the government-sponsored search for a Northwest Passage opened up opportunities for other players to take center stage. Lack of funds for exploratory missions to the Arctic created a similar situation in the Danish context. The case studies in this chapter examine the function, focus, practice, and representation of Arctic science in expeditions funded by three different types of organizers in the 1830s: private patrons, trade companies, and religious missions.

Within the context of rising disillusionment with the quest for the Northwest Passage, the abolishment of the Board of Longitude and related Parliamentary rewards, and the changing character of the Arctic trading companies, the nature of Arctic narratives shifted. Expeditions organized by the British and Danish governments had science as their stated secondary priority, but other organizers did not necessarily follow the same model. Undoubtedly financial concerns played a factor in the organization of the earlier missions, but their importance reached another level when organized by, for example, a trading company. When the financial aspects of scientific exploration took center stage, Arctic science changed in significant ways. We saw in the previous chapter that the type of scientific knowledge produced during these expeditions was not always what the organizers had hoped they would accomplish. This chapter further expands upon this theme and shows the multiple ways knowledge about the Arctic was constructed and presented outside of government-sponsored endeavors. It was still the case that science could be utilized to add credibility to the expeditions and narratives, but in the nexus among economic, missionary, geographical, and scientific concerns, the many strategies available for constructing authoritative narrative formats also came with new challenges.

INGLORIOUS STEAM: JOHN ROSS'S SECOND EXPEDITION FOR A NORTHWEST PASSAGE

In 1818 John Ross suffered the embarrassment of mistakenly determining that there was no passageway through Lancaster Sound in Baffin Bay. What was worse, he claimed to have seen a mountain range, the "Croker Mountains," which soon after turned out to have been a mirage. The British Royal Navy carried out twelve Arctic expeditions in the period between 1818 and 1837, but Ross's falling-out with John Barrow meant that the Royal Navy was not interested in his service again. But there were other ways of financing an exploratory expedition. Ross succeeded in convincing the gin magnate Felix Booth to pay for an expedition in search of a Northwest Passage between 1829 and 1833. Ross was convinced that the use of steam vessels would be an advantage in the Arctic. Steamboats could make progress against the wind or in calm weather and push through bay ice—or so Ross had hoped. His ship, the *Victory*, was adapted with a steam engine. This was the first attempt at using steam in Arctic navigation—and it failed spectacularly. As Barrow anonymously wrote in the *Quarterly Review*, fitting the *Victory* as a steamer had been a disastrous idea, making it "the very worst description of a vessel to navigate among ice—and with engines, in the present case, the most miserable that can be imagined."[2] Ross's published narrative from the expedition initiated a controversy over the use of steam, which affected both the perception of his scientific persona and the future of exploratory missions to the Arctic.

In April 1827 a letter on the utility of steam navigation was published in *Blackwood's Magazine*, signed by "Captains R.N. Edinburgh 1827." The anonymous author of this letter was Ross.[3] The same ideas were put forward in *A Treatise on Navigation by Steam* (1828), which was dedicated to and supported by the lord high admiral of Great Britain. Ross addressed what he saw as the advantages of introducing steam, as well as "the general prejudice against innovations" in the naval fleet.[4] It was understandable, Ross noted, that the Admiralty was reluctant to change what had worked so well. Yet he was convinced that the use of steam engines in navigation could revolutionize the British Navy. Ross's second venture was no more successful in finding a passageway through to the Pacific than his first attempt, but the expedition succeeded in something else, namely the crew's survival in the Arctic for an extended period of time. Ross had not intended for the expedition to last four years, and a key reason for their survival, as with, for example, Franklin's expeditions examined in the previous chapter, was his adaptation of Indigenous techniques. Aside from surviving, the expedition succeeded in surveying large stretches of land and made many scientific observations. Notably, Ross's nephew James Clark Ross was the first European to reach the North Magnetic Pole.

The reviews of Ross's narrative did not comment in detail on the *Victory*'s steam engine—aside from noting that the venture had failed—but this became a particularly controversial subject as Ross blamed the failure of his expedition on the engineers. The development of steam vessels was transformative for the nature of British imperial expansion. The historian Daniel Headrick has shown how the introduction of steamboats changed the balance of power in Calcutta when the East India Company "inaugurated a new kind of war: river warfare." During the Opium War, when Britain sought control over trade in China, steamboats such as the famous *Nemesis* were a key tool. But in the late 1820s the use of steam vessels for Arctic explorations was not an idea readily adopted by the Admiralty. As the historian Maurice Ross has written, around 1830 "the value of steam vessels to tow ships of war out of harbour in contrary winds was recognized by the lords of the Admiralty, but that is as far as they would go."[5] Ross proposed his idea of an expedition with a steam vessel to the Admiralty twice, in 1827 and 1828, but was rejected both times. Instead, he appealed to Booth, whom he described as "an old and intimate friend," but who at first "declined embarking in what might be deemed, by others, a mere mercantile speculation."[6] Booth originally rejected Ross's proposal, as he supposedly did not want it to appear as though he was looking to get part of the £20,000 reward for finding the Northwest Passage. It was not until after the Board of Longitude was abolished and the Parliamentary reward was repealed that he agreed to sponsor the expedition. Booth also saw the benefits of steam.

The business partners and engineers John Ericsson (1803–1889) and John Braithwaite (1797–1870) provided the engine.[7] The engine used for the *Victory* consisted of two high-pressure boilers, which had recently been patented. It was designed so that it could reuse freshwater for the boiler, which would save freshwater storage, as well as fuel, by taking advantage of water condensation. The engine proved problematic. Ross was eager to point out that the problem his expedition had encountered rested with this particular engine and not the general principle of steam engines: "In blaming the execution and workmanship of this engine, I must however do justice to the principle, which was judicious, and, under a careful execution, might have rendered this machinery of great service to us on many occasions which occurred in our voyage."[8] Ross went to great lengths to discredit the manufacturers of the engine in his narrative. The dispute between the two engineers and Ross was so bitter that Booth supposedly had to stop Ross and Ericsson from having a duel.[9] The same year that Ross's narrative was published, Braithwaite published *Supplement to Captain Sir John Ross's Narrative of a Second Voyage in the Victory, in Search of a North-west Containing the Suppressed Facts Necessary to a Proper Understanding of the Causes of the Failure of the Steam Machinery of the Victory, and a Just Appreciation of Captain Sir John Ross's Character as an Officer and a Man of Science* (1835).[10] As the long title indicates, the one-shilling pamphlet was published as a rejection of Ross's attempt to place the blame for the failures of his expeditions on Braithwaite and Ericsson. In his narrative, Ross described the steps they had taken to amend the engine, which he lamented did little to better it, and placed the blame for the engine's malfunction squarely on "the constructers of our execrable machinery, Messr. Braithwaite and Erickson." The boilers leaked, and when they were leaking the forcing pump had to be constantly manned to keep the engine going, and "was to be a cause of hourly torment and vexation to us for many weeks, was at length to lead to the abandonment of one of our chief homes, in addition to all the waste of time and money, consequent on the grossly negligent conduct of our engine-makers." According to his narrative, they attempted to modify "the evil inflicted on us by the discreditable conduct of our engine manufactures" early on in the expedition.[11] His narrative's message was that a better engine would have changed the trajectory of the expedition.

When Ross and Booth commissioned the engine from them, the engineers were not told about the real purpose of the ship. Both Ross and Booth wanted to keep their preparations for the venture a secret, because Ross feared getting scooped and Booth did not want to be revealed as the expedition's patron. The engine had therefore not been designed with the explicit purpose of an Arctic exploration and could not withstand the Arctic climate. This was one of Braithwaite's key counterarguments to Ross's accusations of incompetency. According to him, Ross had led them to believe that the steam vessel was intended as an experiment, for war purposes. They had readily agreed to provide the engine; as Braithwaite noted, "I reasonably anticipated that through him we had as good an opportunity as could be desired of practically testing the worth of that improvement of which he thought so highly." Braithwaite emphasized the experimental nature of the engine several times: "In *experimenting*, complication is seldom regarded, since the intention is merely to ascertain facts and results for guidance in practice."[12] The engine had never been intended for use in an exploratory mission in the Arctic, he argued, but as an experimental steam vessel in English waters. Ross responded to Braithwaite in a short eight-page pamphlet, *Explanation and Answers to Mr John Braithwaite's Supplement to Captain Sir John Ross's Narrative of a Second Voyage in the Victory, in Search of a North-West Passage* (1835). While Braithwaite's pamphlet had been inexpensive, Ross's was even more so, as it could be obtained free of charge from his publication office. According to Ross, Braithwaite had not characterized the steam engine as experimental, but as "*fully tried*, and fit for any service."[13] The secret of the expedition, Ross

FIGURE 2.1. John Ross included this illustration in his narrative, showing his ship, the Victory, stopped by ice. *Source*: John Ross, *Narrative of a Second Voyage*, 1835, 175. Courtesy of the Scott Polar Research Institute.

argued, was kept at Booth's request, but there should have been no problems with the engine had it been made to the high standard of Braithwaite and Ericsson's claims. Ross further accused Braithwaite of changing the design and making the boilers out of iron rather than copper, as had originally been agreed.

When Ross's expedition passed through Baffin Bay and reached Fury Beach on August 13, 1829, they found William Parry's abandoned ship and supplies. Parry had led an expedition between 1824 and 1825 to Prince Regent Inlet, where he at that time believed the entrance to the Northwest Passage was located. Because there was more ice than they had expected, his party wintered at Port Bowen in Prince Regent Inlet. In the summer Parry's ship *Fury* was badly damaged and abandoned on Fury Beach with its stores, which were later taken advantage of by other explorers, including Ross. By the end of September Ross "considered that all hope of making any farther progress this season was at an end."[14] He decided to dismantle the broken engine, and take it out of the *Victory* to open up space inside the vessel: "But thus rendering us no service, the engine was not merely useless: it was a serious encumbrance; since it occupied, with its fuel, two-thirds of our tonnage, in weight and measurement. . . . As the engine, moreover, had been considered the essential moving power in the original arrangement of the vessel, the masting, and sailing had been reduced accordingly, since it was presumed that the sails would only be required in stormy weather; so that, in fact, she was almost a jury rigged ship."[15] The engine that Ross had thought could help them push through ice was thus discarded.

In his treatise Ross had noted that all navigators would need to be familiar with the science of steam when steamships were introduced. It is clear from his descriptions of steam engines that he considered himself knowledgeable on the subject. Braithwaite was not convinced of Ross's expertise and sarcastically referred to the author of *A Treatise on Navigation by Steam* as the only one "who will not admit that there is no difference whatever between the common paddle-wheel and the one to which Captain Ross attributed properties at variance with the most simple physical laws—laws well understood even by those who had no pretensions to be thought scientific." The many problems Ross had with the *Victory*, in particular with the paddle wheels, Braithwaite argued, were due to Ross's own errors in calculating the flotation of the ship and because he kept the objective of the *Victory* a secret. Ross countered by arguing that the paddle wheels were in fact immersed properly, and before the boiler broke that the *Victory* had sailed at a rate of six miles per hour—therefore the mistakes in calculations all belonged to Braithwaite and Ericsson. According to Ross, Braithwaite had explained the want of speed en route to Woolwich with the deep immersion of the paddle wheels. When Ross suggested to move the storages on board the hulk, Braithwaite reportedly said no, "undoubtedly because it might lead to my withholding the last payment."[16] Braithwaite concluded his pamphlet by appealing to the public to pass judgment on Ross and attacking his credibility: "I confidently appeal to the whole world whether Captain Ross has not calumniated the makers of his engines in ascribing to them the failure of his steam-ship; and whether it be not the fact that Captain Ross has slandered them, in order to divert attention from his own errors, his own blunders, and from the disgraceful ignorance and incompetency in which all these errors and blunders originated."[17] In this way Braithwaite's criticisms extended from addressing Ross's characterization of the steam engine to his scientific credibility, just as the reviews of Ross's narrative had done before. In contrast with the reviews of Ross's narrative, the notices of Braithwaite's response to him was described in positive terms. The *Monthly Magazine* noted that it carried "conviction with its undoubted veracity." The *Literary Gazette* agreed with Braithwaite that it was fully the fault of Ross, and Ross alone, that the steamer had failed.[18]

As an expedition organized outside the remit of the British government, Ross's venture shows the tension between surveying for the sake of scientific advancement and national glory on the one hand, and financial remuneration on the other. While aspects of the expedition itself could be framed as successful, Ross's published narrative was less so. In many ways it was a repeat of his narrative from the 1818 expedition, and again the glory and praise Ross received upon his return was short-lived. He and his crew, except for the three unfortunates who had died during the voyage, returned to England and were celebrated as heroes. Their long absence had generated a significant amount of attention, so much so that a rescue mission led by George Back had been planned. Ross was knighted on December 24, 1834, was made an honorary citizen of several cities, including London, and received many prizes and medals. He was sent more than four thousand letters of congratulations. Even a panorama that celebrated the expedition was exhibited at Leicester Square.[19] Ross finally received the glory and praise he thought had been wrongly denied him after his 1818 voyage. But in many ways he was his own worst enemy, and just like his 1819 narrative, *Narrative of a Second Voyage* tarnished his credibility.

Narrative of a Second Voyage was written in a day-to-day journal format that emphasized his firsthand observations and experiences. However, it included many value judgments, from very positive self-evaluations to less flattering portrayals of others. His harsh criticisms were not directed solely at the engineers. The introduction gave a retrospective account of his 1818 expedition that removed all blame for the lack of geographical results from himself. Ross described the ships during the 1818 expedition as unfit for the purpose, and lamented that he had selected only two of the crew members—his nephew and the purser. Ross wrote that he threw "no blame on the late Admiralty on this account," because the Admiralty had been given poor advice by people hoping for monetary gains from the expedition. While Ross did not directly attribute responsibility to the Admiralty, he did place it on the Admiralty's lack of knowledge about the Arctic. Unsurprisingly, Barrow was not happy with *Narrative of a Second Voyage*.

Barrow's review, published anonymously in the *Quarterly Review* in July 1835, was scathing. Ross's narrative covered 740 pages printed in quarto format; Barrow thought this absurdly long for a voyage where "the incidents were few, and the results are next to nothing." Ross should have been more prudent and published a shorter account in octavo format because it was "enough to set the most resolute reader at defiance." Barrow did not shy away from mentioning the debacle over the Croker Mountains in his review. While addressing the quality of Ross's map, which he found lacking, Barrow also questioned the veracity of his description of the Beaufort Islands which "consist of *three*, and three only—and that the other *five* in the book chart are, like the Croker Mountains, non-entities." Barrow continued by addressing Ross's complaints about the way he was treated over the Croker Mountains. His attempt at explaining away his mistake—making it appear as though he had in fact seen a mountain, just at a different geographical position than where he had believed—was brushed aside by Barrow, who noted that "when a prudent man gets into a scrape, he suffers the memory thereof silently to die away . . . or, which is better, openly avows his error and thus disarm censure."[20]

Another anonymous review appeared in the *Literary Gazette* that likewise described Ross's accomplishments and persona in very negative terms. The review went so far as to make a long, and highly sarcastic, list of ways in which it was possible to obtain orders like Ross, ranging from "always keep yourself in the eye of the public" to "placard every wall, hole, and corner" with notices about your achievements. This was also the sentiment in a review that appeared in *Chambers's Edinburgh Journal*, which argued that in spite of all the "fuss" made about the expedition, it "has produced no result of the least value." About the 1818 expedition, the *Literary Gazette* reviewer wrote that "the worthy Captain goes over the grounds of his former voyage; and, as seems to be his usual practice, throws blame about him pretty freely."[21] Both the *Literary Gazette* and *Chambers's Edinburgh Journal* chastised Ross for being ungrateful, as he had received a large grant from the government in addition to the income from subscriptions to his narrative, which had been advertised in public meetings. It was therefore in bad taste for Ross to complain about his income from the expedition. A lonely positive review appeared in the *Edinburgh Review*, written anonymously by the natural philosopher Sir David Brewster (1781–1868). Brewster was particularly interested in Arctic exploration and firmly believed it was a worthwhile pursuit. According to the historian Janice Cavell, Brewster's vision of Arctic exploration was, as were many others', shaped by a dual influence of Romanticism and Christianity. Brewster praised both Ross and Booth for continuing the search for a Northwest Passage at a time when "the zeal of the Government sank into apathy, and, like children tired of their toys, they broke in pieces and trampled under foot the mechanism with which they had been so agreeably occupied."[22] While Ross had made mistakes, Brewster argued, these were insignificant compared to the advances made by his expedition and were worth celebrating.

Ross had attempted to prioritize science the same way the Royal Navy had during their expeditions, but he also had to earn a living. Because his venture was privately funded, it relied on the goodwill of the Admiralty to pay its crew for the additional years they spent stuck in the Arctic; Ross also requested additional financial remuneration for himself. Judging from the correspondence between Barrow and Ross, Barrow had initially been somewhat pleased with the expedition. In fact, Barrow agreed to pay Ross's crew their salary for the extra years they had been gone. That his venture had been a private one shows the support Ross and his crew enjoyed in the initial period after their return. This goodwill did not last. In an anonymous review, Barrow described the Admiralty's decision to financially compensate the crew as a way to protect them against Ross's poor decision making:

> On the return of the party from this ill-fated expedition, Captain Ross addressed two letters to the Secretary of the Admiralty—the one giving a summary of his proceedings, and the other stating his utter inability to fulfill the engagements he had entered into with his crew, and praying their Lordships to afford him the means of discharging obligations of so sacred a character. That he had no claim whatever on the public for an ill-prepared, ill-concerted, and (we may add) ill-executed undertaking, wholly of a private nature, will not be denied; and the wealthy individual at whose expense the ship was fitted out, and who made or sanctioned the "sacred" engagements with the men, was the proper quarter to which application should have been made.[23]

While the crew deserved and received a swift decision by the Admiralty to receive pay, with regard to Ross "no such haste was required." While he had been awarded £5,000 upon his return, there was "not a syllable, throughout his 740 pages . . . to manifest the least feeling of gratitude, or sense of obligation."[24]

Ross's focus on his financial situation, as well as his continued rejection of blame with regard to both the Croker Mountains incident and the failure of the Victory's steam engine, was incompatible with the perceived persona of a British Arctic explorer. Ross' request for additional personal economic rewards played a part in the swift destruction of his initially rebuilt public reputation. It is clear from the reception of his narrative that Arctic exploration was to be done for geographical, scientific, and national advancement, not for financial gain or pride. Reviewers chastised his and his publisher's strategy for maximizing financial gain from his narrative. The tension between the way Ross attempted to portray himself and how he was actually perceived reveals the delicate construction of scientific authority, objectivity, and trustworthiness in the Arctic. Ross had been unable to secure the command of another expedition with the Royal Navy, so the privately funded expedition was an opportunity for him to reinstate himself as a heroic Arctic explorer. His published narrative, however, quickly destroyed this image in much the same way as it had done in 1818. Barrow clearly sought to fully destroy any credibility Ross still enjoyed: his narrative was guilty of "gross misrepresentation," his persona was that of a "vain and jealous man," and he was "utterly incompetent to conduct an arduous naval enterprise for discovery to a successful termination." His expedition and his charts of the coastline were useless because Ross was an untrustworthy observer, Barrow argued, noting "the value of hydrography consists entirely in its fidelity."[25]

MISSIONARY NARRATIVES IN GREENLAND

Privately organized ventures such as Ross's were one way in which would-be explorers could fulfill their ambitions outside of government organizations. Another type of Arctic traveler was the missionary. In this section, I examine how the work of Danish missionaries who settled in Greenland for longer periods of time offers insight into how issues of trustworthiness and authority intersected with economics and religious expansionism. Although individual missionaries may have undertaken some travel and some exploration, they typically lived and worked in the same location for extended periods. Though they were not explorers in the traditional sense, many were engaged in knowledge production and published accounts of their lives in the Arctic. Missionaries were also active practitioners of imperial science, and, as Sujit Sivasundaram in particular has shown, the observation, collection, and signification of nature was a key bridge between scientific exploration and their evangelical mission.[26]

This section examines two accounts by Danish settlers in Greenland: a short two-part publication by an anonymous missionary's wife, "Udtog af en Dansk Dames Dagbog, Ført i Grønland 1837–1838" (Extracts of a Danish lady's diary, kept in Greenland 1837–1838), published in the journal *Læsefrugter*. (As the missionary's wife published her account anonymously and her identity is unknown, I will refer to her as "anonymous missionary" throughout this chapter.) The second narrative, *Syv Aar i Nordgrönland* (Seven years in northern Greenland), was published as a book by the theologian and missionary Johan Christian Wilhelm Funch (1801/1802–1867), who spent seven years in Greenland, living first in Illulissat (Jakobshavn) and then Uummannaq (Umanaq).[27]

Missionaries in Greenland worked closely with the KGH, and the two narratives reveal the ambivalent relationship among commerce, religion, and science in the Arctic.

Accounts from missionaries who settled in a semipermanent way in Greenland offer a different window into life and science in the Arctic. There is a large body of scholarship on the relationships among imperial expansion, trade, science and technology, and missionary activities. In his now classic text, Winfried Baumgart pointed out this problematic relationship, showing that commercial activities could stifle missionary goals, and vice versa. Similarly, Catherine Hall has emphasized that the relationship between missionaries and the empire was not straightforward. While Hall's focus is on the role of nonconformists, particularly the Baptist missionary movement in the British imperial involvement in Jamaica, her analytical points can be usefully extended to a study of Danish missionaries in Greenland. Hall argues that missionaries and planters in Jamaica were united in the belief that British culture was superior to Jamaican, linked to their "civilizing mission"—the view that they had a "responsibility to civilise others, to win 'heathens' for Christ." The civilizing mission was not unique to the British Empire; being a general feature of European powers. As Michael Mann has noted "the concept of the *mission civilisatrice* was used above all for the self-legitimation of colonial rule."[28] That is, missionary activity was also a form of colonialism in itself, distinct from but functioning in close relation to geopolitical annexation.

During the 1830s British Christian missionaries, generally speaking, functioned within a universal family narrative,[29] which included the belief that racial differences could be explained by culture and climate, as well as the belief in a patriarchal family order with white men ruling the metaphorical family. By this reasoning, Indigenous peoples were conceptualized as children and the white man as their father. Funch's narrative was deeply embedded in this rhetoric of the civilizing mission, and he argued that "anyone who cares for Greenland and knows about the conditions in the country, would certainly wish that as long as its inhabitants are children, that state guardianship must remain. When Greenlanders at some point reach the age of majority, then let them enjoy all the benefits of their country, as they will then understand how to use them."[30] This view permeated Funch's descriptions of everything related to Greenland. He considered Greenlanders to be childlike and in need of parental guidance, both spiritually and governmentally. The "parents" were, of course, the KGH and the Christian mission under the guidance of King Christian VIII, to whom the narrative was dedicated. According to Funch, Greenlanders had been better off since the KGH and the Christian mission had arrived.

Funch's stated objective in writing his narrative was to provide a useful account of life in Greenland, including descriptions of the environment, focusing primarily on the area around Uummannaq where he lived for the majority of his stay there. To fulfill this goal, the section "Naturbeskaffenhed" centered on subjects such as birds of prey, dangerous wildlife, farm animals, sources of fuel, and the relative unsuitability of the soil for farming.[31] In comparison with, for example, Graah's narrative, Funch's descriptions were limited by his lack of training in natural history. Although Funch collected specimens during his stay in Greenland, the narrative did not contain a list of details about them, either in an appendix or throughout the book. The only consistent appearance of this type of detail was in the section "Naturbeskaffenhed," where he included more detailed observations related to natural history, with the Latin names for a few plants and animals. Funch's descriptions were markedly less detailed than what we see in Arctic narratives from more formal exploratory expeditions. Having graduated from Copenhagen University with a degree in theology in 1824, Funch moved with his wife Isidora Sophie Funch (b. 1806) to Greenland in 1830, where they resided until they returned to Denmark in 1837.[32] There is no evidence to suggest that he had received any formal education in any scientific subjects.

The majority of *Syv Aar i Nordgrönland* was concerned with the way of life in Greenland, ethnographic observations of Inuit, the KGH, the Christian mission, and natural history. The narrative was relatively brief, 128 pages divided into thirty-eight sections. All aspects of *Syv Aar i Nordgrönland*, including the scientific observations, were framed in religious language. It would, however, be a mistake to assume that this rendered it a niche publication targeted only for other religiously inclined readers. On the contrary, missionary science was part of a diverse body of nineteenth-century popular scientific literature, both in Britain and Denmark, where theological themes were pervasive. The scientific observations of missionaries were also used by more elite scientific researchers.[33] Funch sent papers back to Denmark that were published in the journals of learned societies. His writing was not limited to the religious situation in Greenland or other observations related to ethnography. As was typical for observations made in situ, Funch emphasized that his narrative was a strictly personal account, having only recorded what he observed firsthand. Funch collected specimens of fish, animals, and plants, and some of specimens were for example given to the Danish zoologist Johannes Reinhardt (1778–1845). These became part of Reinhardt's "Ichyologiske Bidrag til den Grönlandske Fauna."[34]

Syv Aar i Nordgrönland was not extensively reviewed, nor was it translated into other languages. This is also the case for the account written by the anonymous missionary. Even so, both are still significant historical documents as they show the tensions among and the interconnectedness of trade, imperialism, religion, and science in the Arctic. Spiritual expansionism was different from state-led imperialism, but the two worked hand in hand in Greenland. Funch's claim that the Danish imperial presence was bettering the lives of colonized Greenlanders was a way to legitimize his own missionary project. It also factored in the way he portrayed himself as a trustworthy observer of life and nature, as they related to religious, political, and scientific subjects. In the section on trade ("Handelen"), Funch's narrative touched upon one of the darkest aspects of Danish colonialism, namely the practice of taking Greenlandic children away from their families and raising them in Denmark under state guardianship (*Formynderskab*). These children were to learn Danish and become educated in various trades. While it may appear surprising that Funch would discuss it as part of his account of the KGH, his views of trade in Greenland, as well as the Christian civilizing mission, were shaped by the idea that Inuit were unable to take care of themselves.

Children and education played a key role in colonial and missionary projects. This was not an idea new to the nineteenth century, or exclusive to the Arctic—"give me a child until he is seven and I will show you the man" is an oft-quoted phrase attributed to Ignatius of Loyola, a cofounder of the Society of Jesus (Jesuits). Schooling, and the control of education, has been used as a tool to reinforce and transform societies through their children. The exploitative practice of removing Inuit children from their families and placing them in homes in Denmark continued through the twentieth century. The author Tine Bryld famously collected the stories of twenty-two children who were brought to Denmark in 1951 which give a chilling insight into the relationship between Denmark and Greenland. This was done "with the best intentions," as Bryld writes.[35] This practice, both in in the 1830s and 1950s, was rooted in ideas of social improvement and civilization. It was not unique to Danish imperialism either, but had parallels throughout the British Empire, as well as post-confederation Canada.[36] Missionaries held a significant role in these civilizing projects aimed at shaping the identity of children into less Indigenous and more European subjects for the empire. From this perspective, accounts such as Funch's are noteworthy in two ways. First, the narrative offers modern readers a window into the practices of missionaries and their attitudes toward the Indigenous members of their congregation. Second, it also functioned as evidence for its contemporary audience of the missionaries' presence, as well as that of the KGH, on the lives and morality of Indigenous peoples in the Arctic.

Whereas nineteenth-century Arctic explorers were mostly male—not including the Indigenous women encountered or employed for assistance and company during the expeditions—settlers in the Arctic were both male and female. The historian Mary Louise Pratt has argued that there were few female travel writers because "seeing" was equated with a masculine desire to possess. Pratt notes that "while women writers were authorized to produce novels, their access to travel writing seems to have remained even more limited than their access to travel itself, at least when it came to leaving Europe." Similarly, Sherrill Grace has emphasized the masculine paradigm of Arctic narratives and argued that "human agency (and with it power, freedom, individuality) has been constructed in northern narratives, and elsewhere, as exclusively male, aggressively heterosexual, and masculinist."[37] Yet, as Pratt's analysis further shows, "imperial eyes" could belong to both male and female travelers, who afforded different perspectives on the land and peoples encountered. While rarer, textual accounts of the Arctic written by female authors do exist, though they have received significantly less attention than those by their male counterparts. One example was the two-part serial by the anonymous missionary, which did not refer to the author's specific identify or location in Greenland.

While there is no information on the background of the anonymous missionary, she appears to have had no substantial training in theology, linguistics, or natural history. Written in the 1830s, "Udtog af en Dansk Dames Dagbog" is stylistically awkward, especially when compared to the literary trends in Britain at the time. As a travel narrative, it was written in a diary format but was largely void of the types of dramatic flair so prevalent in other Arctic narratives of this period, including Funch's. Bernard Lightman has shown how the maternal tradition, a style of writing that adopted the narrative voice of a mother figure, was popular among female writers in the first part of the nineteenth century.[38] The maternal tradition, or the "familiar format," was not well suited for attracting the emerging mass audience of mid-nineteenth-century Britain that comprised men and women of all ages. As such women popularizers of science began experimenting with other narrative formats. With its emphasis on the home and a narrative voice explicitly gendered female, "Udtog af en Dansk Dames Dagbog" has strong parallels to the maternal tradition described by Lightman, and yet it was not written just for women and children.

Læsefrugter, the journal that published the account, was geared toward a broad audience and was an exceptionally popular literary publication.[39] The history of the print press is highly specific to each country, and even to individual cities. In Denmark the cheaper forms of popular science publications and science lectures were not launched until the last decades of the nineteenth century.[40] The explosion of the cheap periodical press and the mass audience in Britain did not happen until the end of the nineteenth century in Denmark. While British publications were translated into Danish, and vice versa, it should not be surprising that narrative formats varied in the two contexts. This was also the case in British North America.

The narrative by the anonymous missionary is significant in that it problematizes the persona of the Arctic explorer. Arctic explorations and their associated glory, danger, and discovery were gendered in opposition to the homebound, passive, and feminine. The anonymous missionary did not establish a feminine version of the heroic Arctic explorer, but utilized the diary format of the travel narrative to create an authoritative, yet passive, narrative. The narrative provides a small window into life in the Arctic for women settlers. She included details such as the weather and temperature, religious services, food resources, and trade. In contrast with Funch's account, the anonymous missionary observed the Arctic almost exclusively from her house and church. Her recorded interactions with Greenlanders were limited to when visitors came to trade or in the context of religious services. All of these encounters were described in terms of fear, uneasiness, and racial bias. In one instance, an unnamed man visited to request a prayer book with songs while she was home alone. She appears to have had limited linguistic abilities besides her native Danish, and she was unable to understand the man's request. To make himself understood, he sang parts of a psalm from the desired book. This, she described, made her "very fearful, and thought that he was insane" until she recognized the tune of the song. The anonymous missionary's narrative was also embedded in the rhetoric of the civilizing mission. She described Greenlanders as childlike, unable to take care of themselves or plan for the future. When the hunting season failed, it was "sad for the Greenlanders," but we are made to understand that it was just as sad for her, as her starving community members visited her family to trade small items for food and coffee.[41] The lack of sympathy for the plight of Greenlanders during a time of food shortage was explicitly linked to her disdain for what she perceived as their unwillingness to save and plan for the future. By contrast, she described the modesty of her household economy and how she had saved and treasured a small bag of potatoes imported from Denmark. With reference to the nineteenth-century St. Lawrence Valley in Lower Canada, the historian Colin Coates has shown that imperial visions of the land were conceptualized in old-world terms.[42] In the anonymous missionary's short narrative, she firmly established her household in Greenland as an extension of her home in Denmark, in contrast to the practices of Indigenous families.

While her descriptions indicate that the anonymous missionary was unwilling to share her food supply, it appears from her narrative that her household may have experienced a food shortage as well, related to the KGH. Funch's narrative, more so than hers, reveals that the relationship between missionaries and the trading company was an ambivalent one, especially when it came to finances and food supplies. The KGH was obliged to support missionaries with sustenance, boats for transportation, and maintenance of mission houses. In practice, this situation, which Funch described as a "dependent relationship with the merchant," was more complicated. The extent to which the traders working for the KGH were made to support the missionaries was contingent on the availability of resources, the determination of which was left to their own discretion. As Funch further noted, if a missionary "demands food . . . they would only have to answer that they did not have any. If he demands a vessel it could easily be in use, and there is thus many ways for the merchant to harass the missionary."[43]

The KGH enjoyed a trade monopoly in Greenland. Though they were also allowed to sell a limited amount of goods, Funch advised other missionaries to tread carefully when engaging in trade on their own, as their financial gain necessarily would be cutting into the company's trade, which could upset their relationship. While Funch supported the monopoly trade in Greenland against the possibility of privatization, he also criticized some of the KGH's practices. Funch's ethnographic descriptions were in part framed to disprove what he considered to be false beliefs about Greenland and Inuit. In response to this, Funch countered the KGH's attempts to force Inuit to remain in the villages during the summer. The KGH was unhappy with the loss of trade during the warmer period, but Funch argued it was unkind to forcefully stop Inuit from traveling to hunt during the summer. Not only did the annual hunt for reindeer provide much enjoyment, the meat sustained the village during the winter and maintained their independence.[44] On the other hand, Funch also argued that it was not unfair that the KGH paid their Indigenous traders significantly less for their products than what they charged for them in Denmark, and conversely sold Danish products to Greenlanders at a marked-up price. Yet, just as the anonymous missionary, Funch believed that Inuit were unable to save money and plan for the future. Because of that, the financial practices of the KGH could partially be excused, as "why should people, who are so careless with their money, have a greater part?," Funch asked rhetorically.[45] While the accounts of Funch and the anonymous missionary were not part of ventures organized by the KGH, the trading company still shaped their experiences.

In addition to access to the trade in Denmark, a key concern for settlers was getting fuel for use in heating and cooking. Funch's description of the available fuel reveals the close link between of natural phenomena and economic concerns. While there was some peat (a popular choice of fuel in Denmark), it was present in sparse amounts and could not be used for anything but cooking. The other main fuel source was coal. Funch did not attempt to give the taxonomical name for the types of coal available, but limited his description to include their smell and appearance when burned. While coal had been available only in limited quantities, he believed that there were plenty of possibilities for extracting large amounts of coal in the north of Greenland, which would help "when the ration of coal sent from the homeland did not suffice."[46] Funch argued that it was possible to live a good life in Greenland, even though it was generally characterized by such hardship that could not easily be comprehended in Denmark. Funch's eye for the possibility of economic advances, as well as bettering the quality of life (enjoyed primarily by Danish settlers), is evident throughout the narrative. By contrast, the anonymous missionary's account was a passive one, in that it afforded little to no judgment on what could be done to improve the quality of life. When she described the winter period, she simply noted that Inuit had starved so much that most of their dogs had died from hunger.[47] Although she expressed sadness at this fact, it was clouded by self-pity for having to engage with their tragedy. On the whole, the ethnographic descriptions in her narrative were highly negative. She framed Inuit religious practices around death and illness, their treatment of animals, and their trustworthiness and general morality in a combination of civilizing rhetoric and disdain.

The relationship between the trading company and the missionaries adds another dimension to the tension between economics and Arctic exploration. Funch warned other missionaries not to appear greedy, as this could ruin their relationship with the KGH. Moreover, too much interest in personal financial advancement ran counter to the established persona of a missionary. There are clear parallels with how the reception of Ross's narrative reveals that appearing too interested in financial gain could destroy any attempt at constructing oneself as an objective and trustworthy observer of the Arctic. When Arctic explorations were not government sponsored, costs were a key challenge. But while explorers and missionaries required money to finance their expeditions, they could not be too forthright about actually needing it. Both Funch and the anonymous missionary emphasized their frugality, and this financial disinterest was part of the way they established themselves as authoritative voices on matters pertaining to Greenland, which was just as important for a missionary such as Funch as it had been for Ross. As the next section shows, it was also a significant challenge for explorations organized by the trading companies. The position of the anonymous missionary was different, as her account was gendered female and nameless. At the same time, there are key parallels between the way Funch and the anonymous missionary portrayed themselves in contrast with the Indigenous peoples. As they made a home in the Arctic, they brought with them ideologies and preconceived notions from Denmark, which shaped their observational and narrative practices and to varying degrees was transformed in the contact zone.

THE HBC TAKES CHARGE: THE DEASE-SIMPSON EXPEDITION

Since the amalgamation of the HBC and the NWC, the new company, led by Governor in Chief George Simpson, had supported the Royal Navy in exploratory missions to the Canadian Arctic. The first expedition organized exclusively by the HBC aimed at tracing the unmapped areas of the northern coast. They began in present-day northern Alberta, at Fort Chipewyan to Point Barrow, located in present-day Alaska, and traveled east between Turnagain Point and Fury and Hecla Strait in present-day Nunavut. The HBC chose two men for the expedition: Thomas Simpson (1808–1840) and Peter Warren Dease (1788–1863). The two explorers were praised in the international newspapers as well:

> An extraordinarily important discovery has been made. For two hundred years the dissolution of a geographical problem under the name of the Northwest Passage has been sought in vain. It has now been found! Dease and Simpson are the names of the two English sailors who on August 3rd 1837 were the first to see the southern flowing world ocean. This discovery, of which you can thank the so-called Hudson's

Bay Company which had sent out the expedition, is of the utmost importance, and the names Dease and Simpson have therefore become historically famous.[48]

While the Danish newspaper was too enthusiastic in announcing their discovery of the Northwest Passage, the expedition succeeded in mapping an unprecedented amount of land.

Simpson's account of the expedition, *Narrative of the Discoveries on the North Coast of America, Effected by the Officers of the Hudson's Bay Company, during the Years of 1836–39*, was published posthumously in 1843.[49] As the first expedition organized by the new HBC, it reveals tensions in the power relations between the company and the British metropole (including the Royal Navy), its effect on the nature of Arctic explorations, the portrayal of the explorer and explorations, as well as the Arctic itself. In particular, Simpson's narrative, and its dramatic path to publication, shows the difficulties explorers faced in constructing appropriate identities for themselves as trustworthy observers when participating in expeditions organized outside the remit of the government.

Dease and Simpson both had backgrounds that were well suited for this type of overland expedition. Peter Dease, the leader of the expedition, was the fourth son of Superintendent General of the Western Indians John Dease (ca. 1774–1801) and possibly of mixed Irish and Mohawk descent. When he was only thirteen years old, Dease joined the XY Company and continued to work as a fur trader after its amalgamation with the NWC in 1804. Dease was appointed chief trader in the new HBC in 1821 and participated in Franklin's second expedition, which was examined in the previous chapter. He supported Franklin's expedition by assisting the negotiations between the Tlicho and T'atsaot'ine peoples, a conflict that had hindered the expedition. In addition, he helped Franklin to manage relations with other Indigenous groups and to obtain food and other provisions. Known for his ability to establish good relations with both Indigenous peoples and his direct subordinates, and for his language and travel skills, in 1828 Dease was promoted to chief factor.

Thomas Simpson was a very different type of person. Cousin of the HBC governor in chief George Simpson, Thomas had attended King's College in Aberdeen, where he graduated with a master of arts in 1828. He was awarded the university's Huttonian prize, their highest award for best overall achievement. Originally Simpson had intended to study divinity, but was made an offer from the HBC to become George Simpson's secretary, which he accepted in 1828. His studies in Aberdeen had prepared him well for the scientific aspects of

Arctic expeditions, and he quickly became an excellent traveler. However, while Simpson was a highly skilled explorer and scientific researcher, his personality was said to have been disagreeable and unstable.[50] While Dease did not publish a narrative from the expedition, Simpson prepared his as he traveled south en route to England.

The Dease-Simpson expedition signified a change in direction of HBC-sponsored expeditions (and thus of the British government, as their representative) in two significant ways: they differed from the Royal Navy expeditions by according a lower priority to scientific subjects other than geography, and they successfully adopted Indigenous methods for surviving and traveling in the Arctic. Previous expeditions to the Arctic had generated large contributions to many scientific subjects, and this was expected of Arctic explorations. The achievements of this HBC expedition showcased what the company could accomplish in surveying compared to the Royal Navy. The HBC prioritized the pursuit of science only so far as it could assist the economic goals or social status of the company. Dease and Simpson were ordered to survey much more land than any of the Royal Navy–sponsored expeditions had accomplished before, but they generated only a comparatively small amount of other scientific research. The Dease-Simpson expedition has therefore not received much attention by historians of science. Yet, just as with the Danish missionaries in Greenland, it does provide important insight into the expression of Arctic science and exploration when carried out in the context of tensions among science, economic gain, and sociopolitical status.

The geographical results of the expedition were acknowledged upon its completion, and in 1839 Simpson posthumously received the Royal Geographical Society of London's medal for "advancing, almost to its completion, the solution of the great problem of the configuration of the northern line of the North American continent." His published narrative itself was a scientific document beyond its geographical aspects. Although the focus on science during the expedition had been downgraded, the reviews of Simpson's narrative reveal that it was still seen as a scientific text, as well as an entertaining account of the dangerous life in the Arctic. For example, the *Aberdeen Journal* noted that "its value, *scientifically*, is really great," while the *Monthly Review* considered that the narrative "will be interesting to the general as well as to the scientific reader." The *Examiner* wrote that Simpson "is to be added to the long list of resolute and daring men, who have perished in their ardour for science."[51] For example, the narrative included observations related to magnetism, the aurora borealis, minerals, plant and animal life, and ethnography. As travel narratives were a key evidentiary resource for both researchers and laymen interested in extra-Europeans, descriptions of Indigenous peoples in texts such as Simpson's both assisted ethnographic researchers and informed the general reader. Simpson's description of the behavior and customs of Indigenous peoples, as well as their physical appearance, style of clothing, and methods of traveling and surviving in the Arctic, was part of a broader discourse on the Arctic and its inhabitants.

Simpson's narrative and the briefer accounts of the expedition that were published in the periodical press shaped not only perceptions of Indigenous peoples, but also of the HBC and the fur trade's influence on their lives. Sending out their own expedition was from this perspective a central way for the HBC to control the discussion around their suitability to govern the territories and create a positive image of the company's policies. The HBC submitted an application for renewal of its license in 1837, and it was not by chance that this coincided with the expedition. The company faced serious criticism of their treatment of Indigenous peoples, particularly by the surgeon and cofounding member of the Ethnological Society of London, Richard King (ca. 1811–1876). King had traveled through HBC territory as part of the Back expedition and was very vocal about the company's exploitative practices when he returned to England. The HBC's directors were anxious to bar King from returning to their territories and to counter his portrayal of their treatment of Indigenous peoples.[52] In his narrative, Simpson also addressed the Indigenous peoples' living conditions, lifestyles, and relationship with the HBC, but unsurprisingly gave a much more positive description than King's.

While at Fort Confidence Simpson described the relationship between the Indigenous fur traders and HBC officers as a familial one: "Every circumstance indicates a kindly familiar intercourse; the natural result of which is, that the Indians are attached to the Company's officers, whom in common discourse they style their 'fathers' and their 'brothers.'"[53] By describing HBC officers as parental figures, he was drawing on the same rhetoric utilized in Greenland by Funch and the anonymous missionary. The parent-child rhetoric implied that extra-Europeans benefited from being guided, or rather controlled, by European policies.[54] Simpson's narrative in this respect was highly biased, constructed to portray the HBC in a flattering light. According to him, the HBC wanted "to render the natives comfortable," with food, clothing, and ammunition, even when they did not have the means to purchase it, without putting them in debt. Yet he argued that "the improvidence of the Indian character is an unsurmountable obstacle to its success," which compelled the company to create policies to control their behavior, including the prohibition of liquor. Simpson also recounted how the company had taken the "human precautions" of vaccination of Indigenous peoples against smallpox. All of this, he argued, showed "the Company's humane policy."[55]

British reviewers of his narrative generally accepted Simpson's positive portrayal of the HBC as truthful. For example, the *Quarterly Review* published an anonymous review of *Narrative of a Discovery* which noted that "there is one fact, evidence of which pervades the volume, and which makes us rise from its perusal with peculiar satisfaction: we mean the truly humanising and Christian effect of the operations of the Hudson's Bay Company on the aboriginal tribes." According to the review, the amalgamation of the HBC and NWC had allowed the new company to make positive changes to the living situations of Indigenous peoples, and that "sufficient proofs of this fact appear at the outset of Mr. Simpson's volume, even in his description, though cursory, of the Red River settlement, from which he started for his journey."[56] Similarly, *Chambers's Edinburgh Journal* praised the HBC's policies as represented by Simpson:

> At this stage of Mr Simpson's narrative we are presented with several traits of Indian character, among which may be noticed their insatiable desire for ardent spirits or "fire-water," as they expressively term it; their improvidence and recklessness during seasons of plenty; their passion for the chase, by which they will destroy countless herds of deer and buffalo, leaving the carcasses to bleach on the plains; and their indomitable aversion to pursuits of a fixed and stationary nature. He also notices many of their manners and customs, with which our readers may become acquainted by consulting the recent work of Mr Caitlin; and dwells upon the many humane endeavours of the Company to improve and better their condition.[57]

This description shows the effectiveness of the HBC's rhetorical strategy, as expressed through books such as Simpson's narrative, to frame their activities and policies in North America as a positive influence on the lives of Indigenous peoples. Simpson's descriptions of the HBC was linked to the company's efforts to maintain their authority in the region, but extended into wider debates about the treatment of Indigenous peoples in the British Empire, as well as developments within ethnology and anthropology.

The debate over the treatment of Indigenous peoples did not, of course, end with the Dease-Simpson expedition. Groups such as the Aborigines' Protection Society (APS), which was established in 1837 with King and Thomas Hodgkin

(1798–1866) as central figures, were vocal in their criticisms of the HBC. The APS argued that the company's monopoly on trade was a direct impediment to the well-being of Indigenous peoples, and furthermore that this extended beyond fairly compensating them for goods and labor, as the company's paternalistic and racist policies affected all areas of life and were designed to keep Indigenous peoples in perpetual dependency on it.[58] The Dease-Simpson expedition and its accounts, mainly based on Simpson's narrative, were shaped by and became part of this heated discourse, with scientific, political, economic, and religious implications. The second feature of the expedition relates to the shift in goals between the HBC and the British Royal Navy, namely the company's use of methods that made traveling and wintering in the Arctic more efficient and cost-effective. A key reason why Ross's expedition had succeeded in surviving their unintentionally long stay in the Arctic was due to their adoption of Indigenous methods. The HBC considered the methods used by Royal Navy expeditions inefficient and costly. Because Simpson and Dease were already in Canada, they did not have to spend time or money on transport from England to North America. There were no large and expensive boats involved in the expedition, and the crew was small. However, Simpson's use of Indigenous knowledge did not positively transfer to the way he portrayed them.

While other explorers, including Franklin, certainly exploited the uneven trade value—from the British and Danish point of view—of trinkets such as beads, tin objects, and tobacco, to gain more valuable items such as boats, furs, and ivory, Simpson's description of these interactions clearly reveals his low opinion of the Indigenous population. The tone of Simpson's account makes it clear that the trade value of these items was laughable, and their desire for tobacco and spirits was proof of their bad character and evidence for the necessity of the HBC's patriarchal policies. There is a tension in Simpson's narrative between his stated and implied highly negative views of Indigenous peoples' morality and intelligence and the fact that he and his crew relied on those same individuals for traveling and surviving in the Arctic in order to complete their extensive surveying.

A good example of this tension is recorded in Simpson's account of the journey between Boat Extreme and Point Barrow. It was an arduous trek on foot, so when they encountered a small group of Inuit, they saw an opportunity to acquire umiaks and travel by water instead. This was a much easier way to travel, not least because it saved them carrying their provisions. Simpson described the first sight of the group as filling him and his crew with "inexpressible joy ... but, on our approach the women and children threw themselves into their canoes, and pushed off from the shore. I shouted 'Kabloonan teyma Inueet,' meaning, 'We are white men, friendly to the Esquimaux.'" According to Simpson, this so much eased the tensions that their presence had caused that Inuit "almost overpowered us with caresses." After trading with tobacco, they agreed to lend the expedition an umiak and oars, which were being used for tent poles, "and arranged our strange vessel so well that the ladies were in raptures, declaring us to be genuine Esquimaux, and not poor white men."[59] This point of comparison appears to have been recorded with some pride. Simpson was also given a sketch of the inlet and coastline by one of the women: "I procured, from the most intelligent of the women, a sketch of the inlet before us, and of the coast to the westward, as far as her knowledge extended. She represented the inlet as very deep; that they make many encampments in travelling round it; but that it receives no river. She also drew a bay of some size to the west-ward; and the old man added a long and very narrow projection, covered with tents, which I could not doubt to mean Point Barrow."[60] Simpson used this woman's geographical knowledge, and his wording suggests he actively sought this information out from the "most intelligent of the women."

There is a stark difference in the way someone like Franklin recorded his interaction with the Inuk interpreter Augustus. In his narrative Franklin included details such as personal names, names of tribes, and details of language, in addition to observations of familial relations, customs, and habits. By contrast, while there was in fact much ethnographic detail recorded in Simpson's account, he did not include personal information such as the name of either the woman or the older man who informed him of the geographical features of the coastline. Whereas Franklin utilized the ethnographic aspects of the "geographical gift," as the historian Michael Bravo has termed the process of navigation by Indigenous informant, Simpson was seemingly uninterested in the finer details of who lived in the areas he was traveling through, and did not procure—or at least did not record in his narrative—any such information from the group that lent him the umiak, referring to them only under the general term *Esquimaux*. Because of this, Simpson was unprepared for meeting another group of Indigenous peoples soon after departing in their umiak, and he did not know, nor appeared interested in, for whom the areas he surveyed was home. By contrast Simpson was happy to exploit Indigenous methods for surviving and traveling in the Arctic, but his narrative did not exhibit much care for the people inhabiting the region, unless it was to show the positive influence of the HBC on their morality. Simpson's narrative was clearly shaped by the company's need to create a polished and humanitarian image of itself in order to justify its continued monopoly on trade.

Simpson and Dease experienced the Arctic and interacted with Indigenous peoples firsthand, and as was standard for travel narratives this direct observation gave the account an air of credibility. One event threatened to ruin Simpson's credibility, however: the circumstances surrounding his untimely death. Simpson's narrative included a preface written by his brother Alexander Simpson which gave a biographical sketch, emphasizing the role his brother had played in the expedition over that of Dease, "although Mr. Simpson's name appears only as second or junior officer of the expedition." According to Alexander, his brother had been "the main-spring of the expedition," as he was the only one with scientific training and had surveyed the large area between Great Slave Lake and the Coppermine River on foot without Dease.[61] Alexander had an important reason for emphasizing his brother's skills and role during the expedition. Afterward, on his way back to England, Simpson traveled south toward the Minnesota River with a large party. Traveling ahead of the main party with four men, on June 14, 1840, Simpson shot John Bird and Legros Senior, prior to dying by suicide. Eyewitnesses stated that Simpson had become mentally unstable and thought that the men wanted to kill him. The murder-suicide was extensively discussed in the periodical press, where Simpson was described as a "madman" who suffered from "mental hallucination."[62]

In his preface to his brother's narrative and in his own *The Life and Travels of Thomas Simpson: The Arctic Discoverer* (1845), Alexander emphasized the possibility that Thomas had acted in self-defense and that "the depositions of those who pretend to describe the manner of his death are contradictory in the extreme."[63] This was his way to protect his brother's legacy as a heroic Arctic explorer, give credibility to his narrative and, by extension, the HBC. As it was, Simpson's narrative was well received in Britain. For example, the *Examiner* noted that "the name of Thomas Simpson is to be added to the long list of resolute and daring men, who have perished in their ardour for science, on the scene of their adventure and on the eve of what promised to be their greatest discoveries. This Narrative was found among his papers, and forms a pleasing record of him." The reviews echoed Alexander Simpson's assertion that Simpson was the primary driver behind the achievements of the expedition. The *Critic* was particularly flattering, as it described Simpson's narrative form as "modest": "A more exciting story of adventure, a record more creditable to British courage, humanity, and intelligence, has seldom been offered to the public."[64]

As the mercantile aspects of Arctic explorations were touched upon throughout the narrative itself, the reviews did the same. For example, the *London and Westminster Review* noted: "It was at first thought that few commercial advantages could arise from the discoveries of Messrs Dease and Simpson, but from the nature of the interior, which is intersected by rivers and lakes abounding with fish, and the facilities it possesses for the collection of furs, they are likely to be considerable."[65] While they did not collect or describe a very large amount of natural history specimens during their expedition, Simpson described in detail the potential for further use of these resources in the areas and the process by which the trade was carried out. Inuit were, he stated, often "eager to trade" and "anxious" to trade furs for objects such as shells.[66] More significantly, he also touched upon the import and export of goods. The fur industry was a large transcontinental business industry. In "The Importance of Staple Products" (1930), Harold Adams Innis famously argued that the political, economic, and social development of Canada was shaped by the export of raw materials—that is, staples—to other countries. The staples thesis has been criticized on many level, but is still an influential and useful expression of the economic structure of nineteenth-century Canada, where imported products from England were sold at a high cost.[67] Simpson lamented the fact that it seemed impossible to convince the Indigenous population to become settled farmers and establish Canadian domestic manufacturing of products otherwise produced in Britain. The raw materials should be better utilized in Canada, he argued, as this could "diminish the annual orders from England, and . . . render the people independent."[68] According to Simpson, of course, the organization best suited to support such developments was the HBC.

The role of trading companies within the British Empire remained a heated subject. In 1857, two years before the HBC's grant of the colony of Vancouver Island was due for renewal, the Select Committee on the Hudson's Bay Company reviewed its history as part of a wider discussion on whether such a trading company was suited to govern British colonial land. The company's colonial authority in North America was a concern for the metropole and, as the next chapter shows, was debated extensively both in the British Parliament and in the press.[69] While the Dease-Simpson expedition did not settle the controversies surrounding the HBC, it did showcase what could be accomplished during overland expeditions originating in Canada compared to those sent out from England. In contrast to the expeditions organized by the Royal Navy—and those designed in its image, such as Ross's second expedition—the HBC's expedition did not place a significant emphasis on science. As Simpson himself noted, the company did not even provide his expedition with chronometers.[70]

Simpson's published account shows a tension between his implicit mandate to portray the HBC as a benevolent organization that was perfectly suited to govern the land and its peoples and the familiar conventions for establishing scientific authority. One clear difference between Dease and Simpson's expedition and those organized by the Royal Navy was in the amount of natural history specimens collected. Dease and Simpson did not prioritize this, reflected in the fact that the HBC did not provide them with appropriate scientific equipment. As a trading company, the HBC was concerned with its bottom line, as well as maintaining its monopoly and jurisdiction in its territories. Yet the Dease-Simpson expedition for the most part successfully avoided allegations that their results were compromised by economic infringement. The pressing concern for Alexander Simpson was how to handle the unfortunate way his brother had died, as the allegation of madness was yet another way the expedition's results could be delegitimized. Alexander Simpson defended his brother against these charges by suggesting that his killers had been after the valuable documents he had produced during the expedition. The reviews mentioned in this section suggest that this explanation was not fully believed, yet it did not negatively affect the perception of the expedition itself. Perhaps because Dease and Simpson surveyed an unprecedented amount of land, the reviews do not appear to have considered Simpson's murder-suicide or the lack of collected specimens and experimental results to be other than a minor downfall. In some respects, this speaks to the disappointments with the lack of geographical results from previous expeditions in search of the Northwest Passage.

THE UNEASY RELATIONSHIP BETWEEN SCIENCE AND MONEY IN THE MAKING OF ARCTIC EXPERTISE

While the primary goal of Arctic explorations organized by the British Royal Navy had been navigational, with science as the stated secondary goal, this was not necessarily the direction taken by other types of patrons. As the increasing disillusion with explorations to find the Northwest Passage opened up opportunities for new types of explorers and organizers, the function of scientific discovery and narrative practices in shaping the persona of the authoritative Arctic observer changed. A key difference was the prioritization of formal scientific inquiry and use of expensive equipment, such as chronometers, especially outside of government-organized expeditions in the first half of the nineteenth century.

In some ways, the Ross and Dease-Simpson expeditions are a study in contrasts. One attempted to optimize Arctic exploration by the use of new technologies such as steam engines and large crews; the other adapted Indigenous methods and scaled down the size of the expedition. Their differences were also linked to the emphasis placed on conducting scientific observations and experiments. There were many ways in which scientific research could add credibility to the author of travel literature, yet producing results was not enough. There were certain conventions that furthered trustworthiness. The use of a diary format emphasizing direct observation was a popular strategy for constructing an authoritative narrative format. Another convention was the reference to and summaries of past expeditions, which could be used both to support one argument and discredit others. As the case of Ross showed, however, this technique could be detrimental when used incorrectly. While Barrow initially appeared pleased with the results of Ross's expedition, his perception of Ross changed quickly. Ross effectively blamed everyone but himself for the misfortune of both of his expeditions. His attempts at establishing himself as a more knowledgeable expert on the science of steam than the engineers backfired and came off as prideful and dishonest. As Barrow noted, the charts Ross produced of the coastline were useless because Ross, once again, had proved himself untrustworthy. By reopening the Croker Mountain debacle, blaming the Admiralty for his mistake, and openly attempting to increase his personal financial compensation, Ross wrote his own downfall.

While the narrative format of travel literature did not work well for Ross, it was an important and very effective medium for Dease and Simpson. This expedition, organized by the HBC, did not prioritize scientific discovery to the same extent as those by the British and Danish governments. Yet, the HBC governor in chief saw scientific engagement as a key tool for creating goodwill for the company. It was therefore no coincidence that the Dease-Simpson expedition took place around the time of the HBC's license renewal. Dease and Simpson were largely able to avoid allegations that their expedition was influenced by the company's financial concerns and maintained their personae as trustworthy observers. However, the financial ambitions of the company still shaped the expedition, as was also the case in Greenland with the KGH.

The relationship between missionaries and the trading company in Greenland was one of alternating tension and cooperation. It had a profound influence on the missionary experience, and by extension the knowledge they produced. In particular, Funch's narrative shows a key preoccupation with balancing the portrayal of religion and trade with the missionary's place within the scientific community. This was linked to the construction of his own identity as a suitable person to undertake such work and as a trustworthy source for ethnographic data. The anonymous missionary's short diary provides a unique perspective, being largely void of the drama that was so present in other Arctic narratives. Yet both authors were in Greenland as part of the Christian mission, and their narratives shed light on the relationship among Christianity, science, and imperial expansion. Funch and the anonymous missionary were not commissioned by the scientific societies in Denmark to undertake research, yet their accounts still added to the body of knowledge about the Arctic, in particular in the field of ethnography, including religious practices and linguistics. There is a long tradition of missionaries contributing evidentiary resources to ethnographic research, and this was also the main focus of Funch and the anonymous missionary. Their voices and the Arctic they constructed do not comfortably fit into the rhetoric of the (male) heroic Arctic explorer, and as scientific documents their travel reports differ significantly from those produced by explorers such as Ross and Simpson. The anonymous missionary did not frame herself as a heroic Arctic explorer, but focused on the home, not unlike the maternal tradition that was popular among female writers in Britain, though her account was aimed at a broad, general audience and was not written specifically for women or children. Similarly, Funch's process for establishing himself as an authoritative persona was different than for Ross's or Simpson's, as reflected in their scientific focus and narrative practices.

While exploration for financial gain was certainly also central to the expeditions organized by, for example, the British Royal Navy, the privately funded explorers were more explicitly navigating charges of vested economics interests. This was clearly the case both for the Dease-Simpson expedition (organized by a trading company) and Ross's expedition (organized by a private financier), and was to a certain extend also the case with the Danish missionaries (organized by the mission and working with a trading company). In all instances scientific research was to varying degrees used to legitimize the ventures against charges of bias. These expeditions show the difficulties faced by diverse types of explorers in justifying or defending their scientific and cultural authority. The disunity of Arctic science becomes clearly evident in such a comparison.

CHAPTER 3

The Lost Franklin Expedition and New Opportunities for Arctic Exploration

> The Board of Admiralty by their "effort" virtually declare that the lost Expedition cannot be relieved unless the "Passage" be discovered; we must first discover the "Passage" and then seek out the lost Expedition. To this declaration, my Lord, I cannot assent; for by following out my plan, I can search all that is known of the western land of North Somerset—and be sure that every inch of discovery beyond it is so much good work for the safety of the lost Expedition and for the furtherance of geographical and natural history knowledge.
>
> — Richard King, "The Arctic Expeditions," 1847

In 1848 the British Government sent out three missions in search of the crew from the lost ships HMS *Erebus* and *Terror*. More than thirty expeditions followed.[1] The disappearance of this expedition, led by John Franklin, generated a huge amount of publicity, in part due to the public campaigns to rally support for search missions organized by the captain's widow, Jane Franklin. While funds for search missions were still not flowing freely, the disappearance of the expedition made them much easier to come by than in previous years. But what was the primary purpose of these missions? Officially the goal was finding Franklin and his crew, and everything else came second. But did it? Soon after it became known in England that Robert McClure (1807–1873) had succeeded in transiting the Northwest Passage, Jane Franklin wrote a letter to the first lord of the Admiralty, Sir James Graham (1792–1861), accusing the Admiralty of organizing relief expeditions only as an excuse to survey for the passage. She noted that although the aim of Franklin's expedition had been to locate it, this was not the point of the search missions, and "could I have expected then, that you Sir, and your colleagues, would have taken the earliest opportunity of showing that as soon as the N.W. Passage was discovered, all further interest in the search was at an end?"[2] Reaching a similar conclusion, Richard King's letter in the *Athenaeum* suggested that finding the Northwest Passage was the Admiralty's key motivator. King had been unsuccessful in securing the command of an Arctic expedition after his fallout with the HBC, and his letter was a public request to be given command of an expedition to search for Franklin. King's letter also reveals an important point about the role of science as part of Arctic expeditions in this period. If he was given command of an expedition, King argued, he would shift focus away from finding the Northwest Passage and prioritize finding the lost expedition and enhancing Arctic science.

Perhaps unsurprisingly, given King's track record, his letter did not win him government support. Other opportunists were more successful, yet searching for the lost expedition added clear challenges that had to be navigated. While the attention surrounding the Franklin expedition made funds available from both governments and private patrons, they came attached to a different level of scrutiny than before. As an anonymous author noted in the *Morning Post*, "The whole case is changed from its original character."[3] This chapter does not focus on the Franklin expedition itself; rather, it approaches the lost expedition as a change in the driver behind the organization of Arctic explorations. I examine three expeditions which searched for Franklin to address the question of what happened with Arctic science when the main goal was no longer discovering the Northwest Passage but finding Franklin and his men: the John Rae (1813–1893) and John Richardson expedition between 1848 and 1849, Rae's later discovery of the fate of Franklin's men, and Johan Carl Christian Petersen's (1813–1880) account from Francis Leopold McClintock's (1819–1907) expedition between 1857 and 1859.

The lost Franklin expedition added new challenges that had to be navigated. It was a popular topic in the general periodical press, poems, books, and lectures, in England and beyond. As Roderick Murchison (1792–1871) stated at a meeting of the Royal Geographical Society of London in 1859, "Whilst Sir Robert M'Clure had been worthily rewarded for his intrepid conduct in making a north-west passage, Franklin was the man who, by the self-sacrifice of himself and his brave companions, had previously, by common consent, made *the* north-west passage."[4] Franklin now personified Arctic exploration in Britain. When Rae reported that the lost expedition had resorted to cannibalism, he was effectively deconstructing the image of the heroic British explorer. The ability of Jane Franklin to generate public interest and financial support for continued expeditions shaped them and the representations of the Arctic explorer in the period between the late 1840s and early 1860s. For the explorer this could be a double-edged sword. The goal was not only to find Franklin but also to find him in the right way and to present the mission in a manner that reflected this. In this chapter, I ask the question: What happened with Arctic science when the official main goal was no longer discovering the Northwest Passage, but finding Franklin and his men?

A GENTLEMANLY ARCTIC EXPLORER

Since the end of the Napoleonic Wars, John Barrow had been an important promoter of Arctic exploration. In 1844, at almost eighty years old and finally nearing retirement, Barrow was eager to promote one last expedition in search of the Northwest Passage. The Franklin expedition was Barrow's last opportunity to solve the mystery that had occupied so much of his life, and it is hardly an exaggeration to describe it as the biggest failure of his career. The last Arctic expedition organized by the British Royal Navy prior to this had been led by George Back between 1836 and 1837. The intention was that Back should be gone only one season, so as to avoid wintering in the Arctic. Perhaps unsurprisingly, the plan did not work out, and Back's ship, the *Terror*, froze in. When the expedition finally was able to escape in July 1837, they returned home in a severely ice-damaged ship. While Back's expedition had been unsuccessful, the HBC-organized Dease-Simpson expedition had charted much of the last unknown coastline. Barrow believed that there was a still unmapped coastline between Melville Island and the Bering Strait, and that this could, with the current state of geographical knowledge, readily be charted, and thus the Northwest Passage completed. The lack of results from Back's expedition, however, made it difficult to gather enough support for another venture from the lords of the Admiralty, the sailors, or the general public. When James Clark Ross returned from a three-year expedition to the Antarctic in 1843, Barrow saw an opportunity to capitalize on his success. Barrow submitted a "Proposal for an Attempt to Complete the Discovery of a North-West Passage" in December 1844 to Lord Haddington, first lord of the Admiralty, who accepted it.[5]

In his proposal Barrow drew clear lines between the search for the Northwest Passage and scientific progress: "There is a feeling generally entertained in the several scientific societies, and individuals attached to scientific pursuits, and also among officers of the navy, that the discovery, of a passage from the Atlantic to the Pacific, round the northern coast of North America, ought not to be abandoned, after so much has been done, and so little now remains to be done; and that with our present knowledge no reasonable doubt can be entertained that the accomplishment of so desirable an object is practicable."[6] Arctic explorations, Barrow argued, had contributed to the development of valuable industries such as the cod and whale fisheries. When it came to explorations, Barrow stated that "enlightened minds" knew that the result of "knowledge" was "power."[7] Barrow also invoked the interest in geomagnetism as a reason to continue Arctic explorations, in addition to advances in geography and hydrography. He completed the trinity of arguments by arguing that it was the special privilege and duty of England to complete the search for the Northwest Passage. Money, science, and national power were the reasons Barrow used to promote one final attempt at finding the passage.

The Admiralty first approached James Clark Ross, but he was not interested in another expedition to the Arctic. Franklin volunteered his services, and though he was fifty-nine years old, "the man who ate his boots" was chosen for the expedition.[8] Franklin originally had 134 men with him, including the experienced Arctic and Antarctic sailors Francis Crozier (b. 1796) and James Fitzjames (b. 1813). The ships Terror and Erebus had been reinforced to withstand thick ice and had previously been used by Ross on his Antarctic expedition. The intention was that Franklin's expedition should be completed in one season. Both Erebus and Terror were fitted with steam engines. Rather than custom-building the engine, an old engine from a London & Croydon Railway locomotive was refitted into the ships. Other measures to ensure the success of the expedition were taken: further strengthening the ships and a large store of food supplies, including eight thousand tins of preserves, such as cooked meat and soup, in case they would need to winter in the Arctic. The Terror and Erebus left England on May 19, 1845, and reached Godhavn on Disko Island in Greenland on July 4. They continued through Barrow's Straits and are believed to have wintered at Beechey Island. By autumn of 1847 concerns were growing that, in spite of these precautionary measures, something had happened to Franklin and his crew.

In 1848 the British government sent out three search missions, one overland and two by sea, to optimize the amount of area surveyed. James Ross led an expedition through Lancaster Sound, while William Pullen (1813–1887) went through the Bering Strait. The overland expedition was led by the Orcadian HBC surgeon John Rae and the Scottish naval surgeon John Richardson. The goal of the Rae-Richardson expedition was survey the area between the Mackenzie and Coppermine Rivers and the shores of Victoria and Wollaston Lands in addition to searching for Franklin. Their official instructions from the British Admiralty did not include scientific research. The historian Ted Binnema has rightly pointed out that "perhaps it would seem insensitive to order men to botanize on a rescue mission"[9]; yet Richardson did not lose out on the opportunity to undertake some research. As he had traveled through the Arctic before, he was well aware of the research possibilities there and well prepared to undertake them. Rae and Richardson did not find Franklin, though Richardson left without examining the entire area they had intended. Following Richardson's departure, Rae remained in close correspondence with him and detailed the progress of his surveying.[10] Instead, Richardson focused his narrative, *Arctic Searching Expedition: A Journal of a Boat-Voyage through Rupert's Land and the Arctic Sea, In Search of the Discovery Ships under Command of Sir John Franklin* (1851) on the scientific achievements of the expedition.[11] Richardson's research focus was broad, including ethnographic observations, linguistics, geography, climate, and the natural resources available in the regions surveyed. Although the Admiralty had not explicitly requested them, natural history and their related economic possibilities were the key results of the expedition.

The Rae-Richardson expedition was organized by the British Admiralty, but it was more like the low-budget expeditions organized by the HBC than the large ones usually sent out by the government; their techniques were shaped by Rae's employment with the HBC. Travel was rugged, and they carried very few provisions or scientific instruments, in stark contrast with the lost expedition they hoped to locate.[12] However, their expedition still retained a key feature of those organized by the British Admiralty, namely the focus on gathering extensive information on natural history. Compared to the Dease-Simpson expedition, where they surveyed an impressive amount of land but made very limited scientific observations outside of geography, Rae and Richardson carried out extensive experimentation, cataloging, and collecting. After Richardson left for England, Rae continued to carry out observations, such as variations of the compass, and posted his results to Richardson. There were many similarities between Richardson and Rae. They were both Scottish and trained in Edinburgh. While at the University of Edinburgh, they both attended lectures by the geologist Robert Jameson (1774–1854), professor of natural history. Jameson published a set of instructions in the *Edinburgh Magazine and Literary Miscellany* (1817) that listed artifacts desired for the university museum.[13] The paper included explanations for how to record, collect, and prepare specimens spanning categories such as zoology, ethnography, and mineralogy. Adrian Desmond and John Moore noted in *Darwin's Sacred Cause* (2009) that Jameson's course was "packed with the next generation of travellers: surveyors, civil engineers and army surgeons."[14] Jameson put great emphasis on the importance and role of naturalists going abroad, and his lectures were intended to prepare students for voyages, including Rae and Richardson.

Rae studied medicine in Edinburgh for four years, first at the university, followed by a period at the Surgeon's Hall. He was twenty years old when he entered the service of the HBC as a surgeon on board an HBC supply ship. Rae's first trip was prolonged when the ship was blocked by ice in Hudson Strait, and they had to winter near Moose Factory. Rae was remarkably well suited for life in the Canadian Arctic. He was born in the far-northern Orkney Islands, and Rae's biographers have linked his upbringing there with being so well prepared for his work with the HBC.[15] Rae himself held Orkneymen in high esteem, as evident in his description of a situation where the rough weather of the Arctic was proving troublesome for some of the English men in his party: "I here saw the benefit of the precaution I had taken to have some Orkneymen with me, for it was evident the others (although as good fellows as could possibly be wished) knew nothing about the management of a boat in such weather."[16] Similarly, when Jane Franklin was in contact with Rae about her plans for future search missions, she noted that she would "get part of the crew to be Orkney & Shetland men."[17] The Orkney Islands lie farther north than Fort Churchill, an outpost of the fur trade on the frozen shore of Canada's Hudson Bay and supplied a large number of employees for the HBC. During his stay in Moose Factory, George Simpson, the governor in chief of the HBC territories, offered Rae a position as surgeon and clerk at Moose Bay, which he accepted.

Richardson was also a seasoned Arctic explorer, as well as a surgeon-naturalist. He was a friend of Franklin, and had accompanied him on both the Coppermine expedition and Franklin's second expedition. After studying medicine in Edinburgh, he worked as a surgeon at the Dumfries and Galloway Royal Infirmary before he, as a fellow of the Royal College of Surgeons, secured employment with the British Navy. He was stationed at sea during the Napoleonic Wars, after which he earned his MD from the University of Edinburgh in 1816. Richardson received several honors and awards and was knighted in 1846. His expedition with Rae was his last, and he retired from active duty with the British Navy in 1855. Richardson was a prolific writer and published numerous works from his Arctic explorations, of particular significance being the *Fauna Boreali-Americana*, which he edited. Together the *Fauna Boreali-Americana* and the *Flora Boreali-Americana*, edited in part by the professor of botany William Hooker, are typically identified as establishing the new field of Arctic geographical natural history.[18] These comprehensive accounts of North American fauna and flora were somewhat Humboldtian in their ambition to relate climate to the geographical diffusion and migration of species. The HBC faced a lot of criticism in the period leading up to 1859, when its charter was up for renewal. One of the arguments made by the HBC to maintain its monopoly was that its territories were unfertile and unsuitable for settlement. The only value of the land, the company argued, was in the fur trade. The possibility of establishing farms and securing food products was of no small importance for the possibility of settlements, as having a secure and steady food supply would make it possible to place settlers in more permanent stations. In turn this would strengthen geopolitical claims to resources in the territories. In his narrative Richardson detailed what types of crops and vegetables could flourish at different latitudes, in addition to the availability of game and valuable mineral resources.[19] Richardson's research—which he shared with the British government funding the expedition—played into a large and significant political question: the governing of British North America.

George Bellas Greenough (1778–1856), the first president (1807–1813) of the Geological Society of London (f. 1807), played an important role in shaping its policies. In particular he was committed to establishing a far-reaching network of geological informants. Richardson was not a fellow of the society, but he and Rae were both immersed in an academic atmosphere that encouraged travel for the advancement of natural history. Richardson went into great detail on geology in his narrative, following a modified Wernerian scheme for stratigraphy taught by Jameson in Edinburgh.[20] Abraham Gottlob Werner (1749–1817) was one of the most influential geologists during the early Industrial Revolution. Rachel Laudan's now classic book established Werner's central role in the foundation of modern geology and emphasized the dissemination of his methodological and theoretical preferences throughout Europe in what she termed the "Wernerian radiation." The essence of Werner's teachings was the concept of formations and the formulation of a program of historical geology. As Mott Greene has further argued, above everything else Werner's primary focus was "the empirical establishment of regular successions of strata wherever they appeared and the immediate employment of the knowledge of that succession to serve practical and economic ends."[21]

Jameson, who taught both Rae and Richardson, studied with Werner in 1800 and established the Wernerian Natural History Society in Edinburgh in 1808. Jameson's teaching likely had a great influence on both men. Richardson's narrative shows his focus on geology and efforts to highlight the importance of the mineral resources that were available in the region. His detailed account of the geological features of the surveyed areas highlighted the possibilities for coal extraction, a key focus, as well as other valuable minerals "of far greater value than all the returns which the fur trade can ever yield."[22] Determining the location of those economically important areas was of considerable significance. For example, Richardson described the southern shore of the Mackenzie River as belonging to the Erie division of the New York system, categorized as part of the Silurian system by American geologists and as part of the Devonian (carboniferous) series by their English counterparts. The issue of the Wernerian notion of universal formation of different lithologies (the physical characteristics of a rock or stratigraphic layer) was a central feature in the decadelong controversy in British geology over the Devonian system. Originally used to categorize any rock or fossil found in Devonshire, the meaning of *Devonian* had transformed by the time of English and international geologists reached consensus in the 1840s to denote a fossil, rock, or event that had originated during a specific period. There was a large economic factor related to the theory, such as creating geological maps for finding coal deposits. Furthermore, Richardson believed that there was more iron chromate, a very valuable mineral, in its primitive porphyry form in North America than in England or continental Europe.[23]

The link among the Erie Division, the Silurian, and the Devonian series was not universally accepted in the nineteenth century. For example, the Canadian geologist, paleontologist, and administrator John William Dawson (1820–1899) devoted much research to determining that the Erie Division, or Erian Period, in North America should be the typical region of the Devonian.[24] Geological research on how the strata were classified was fundamentally contingent on the analytical framework of the observer. While Richardson in his previous writings had followed a modified Wernerian framework for analyzing geological data, there was no need to address those issues in *Arctic Searching Expedition*. He did not refer to the Wernerian analytical framework for categorizing strata but still primarily made use of the classical Wernerian approach of valuing physical characteristics of rocks and stratigraphy over specific fossils or other paleontological evidence, following the British geologist William Smith (1769–1839). Yet Richardson also paid attention to fossils, and, for example, noted that the Pentamerus limestone was named from its characteristic fossil.[25] He also included illustrations of fossil plants in his narrative, which was reflective of changes within the broader geological community. Through figures like Richardson and their narratives, the Wernerian approaches were a central part of geology research in the British Empire, including during the search missions.

In addition to surveying for valuable minerals, Richardson noted the types of food resources that were available or could be farmed at different latitudes. While potential financial benefits of the Northwest Passage as a trading route were doubtful, there was another not-insignificant economic motivator to continue explorations of the northern shoreline: the discovery of new fishing grounds. This was linked to Richardson's discussion on climatology. During the expedition, he compiled a comparative table of temperatures. As in the *Fauna Boreali-Americana* and *Flora Boreali-Americana*, Richardson's climatological observations followed a Humboldtian spirit, using Humboldt's concept of isothermal lines introduced in his 1817 work *Des lignes isothermes et de la distribution de la châleur sur le glob.*[26] Richardson's table included the mean annual temperature, isotherms, mean summer temperature, *isothæral*, and mean winter temperature, *isocheimenal*. He also used the meteorologist Heinrich Wilhelm Dove's (1803–1879) temperature tables published in the *Report of the 17th Meeting of the British Association for the Advancement of Science* in 1847.[27]

Humboldt and Dove both drew isothermal maps as a way to visualize weather patterns. Richardson did not include this type of visual representation, but he followed the key Humboldtian ideal, as Dettelbach has described it, of a "universally legible nature." Cataloging the climate in the Arctic, as in other places in the British Empire, was linked to imperial expansion, as it gave indications of where it was possible to settle. Meteorology and state control were closely intertwined in the British imperial context, for example, the British interest in meteorology in India for medical and topographical reasons. For Canada in particular, the changing political situation in the HBC-governed territories was linked to understandings of the Humboltian isotherms. As Suzanne Zeller argued, "Humboldtian science thus heated up support for Canada's annexation of Rupert's Land as natural, perhaps even inevitable."[28] Was it possible to have a flourishing agricultural expansion into Rupert's Land? This was the key question Richardson addressed in his section on climatology.

In addition to the comparative table of temperatures, Richardson included tables for the geographical distribution of plants and the number of species in different zones. From his observations, Richardson came to several conclusions on the suitability for various agricultural choices. He divided North America into five regions, according to the physiognomy of their vegetation: "If we trace any one of these districts northwards, making due allowance for the varying altitude of the country above the sea, we may ascertain the effect of increase of latitude on the vegetation of that meridian; but if we compare one district with another, we must keep in view the climatological fact of the rise of the isothermal lines in proceeding westward."[29] This division helped to account for variations in the presence of vegetation and the further possibilities for cultivation at the same latitude in different places. Significantly, Richardson noted that while there may be fewer species of plants at high latitudes, the number of plants each individual species produces remains the same. In Rupert's Land, governed by the HBC, there was "dense herbaceous vegetation."[30] In many ways mirroring the content of Barrow's proposal from 1844, Richardson's natural history observations were of scientific, economic, and geopolitical significance. Richardson implicitly countered the HBC's arguments that their territories were not suitable for settlement by enumerating the many natural resources in the regions.

The observations Richardson undertook in HBC territories and the conclusions he reached in his narrative were influenced by his understanding that variations in annual temperatures could be meaningfully represented and predicted by means of isothermal lines. On the one hand, he distanced himself from making more explicit value judgments on the suitability of the HBC to govern its territories, writing that "without entering into the question of the chartered rights of the HBC, or the propriety of maintaining a monopoly of the fur trade, it is my firm conviction, founded on the wide-spread disorder I witnessed in times of competition, that the admission of rival companies or independent traders into these northern districts would accelerate the downfall of the native races." On the other hand, Richardson directly questioned the efficacy of HBC governance. In his description of the Osnaboya (Assiniboia) colony, he noted that "the settlement is under the government (it can scarcely be said the control) of a governor, council and recorder, all nominated by the Hudson's Bay Company." Richardson further criticized the ability of the company to even enforce their monopoly against attempts by "half-breed settlers, encouraged by some of the colonial merchants and Roman Catholic priests" to "share the fur trade" with it. The HBC, Richardson argued, "do[es] not seem to possess a force adequate to prevent their eventually succeeding in their object." Richardson also scolded the company for allowing their fur traders to supply Indigenous peoples with alcohol.[31]

Although Richardson held individual HBC officers, including Rae, in high esteem, his portrayal of the company's governance of its territories was not very positive. He also exhibited a highly negative and damaging attitude toward

Indigenous peoples in his narrative, which contained extensive ethnographic observations. During his account of what he termed the "Chepewyan" people, Richardson noted that "they can scarcely be said to esteem truth a virtue." Such value judgments are revealing. Richardson and Rae were interviewing Indigenous peoples about whether they had seen any trace of Franklin. In the section "Interview with Eskimos," he argued that "neither the Eskimos, nor the Dog-rib or Hare Indians, feel the least shame in being detected in falsehood, and invariably practice it, if they think that they can thereby gain any of their petty ends."[32] Richardson's narrative was full of this type of highly derogatory comments toward the Indigenous peoples. As the next section will show, this illustrates the subtle ways the question of who was a trustworthy observer of the Arctic had widespread and unexpected consequences.

Richardson produced an impressive amount of research during the expedition because had largely let Rae undertake the bulk of the search for Franklin—the actual objective of the expedition. Rae and Richardson did not find any tangible evidence of expedition's fate, but Rae continued to inquire after its whereabouts during his later HBC missions. Without speculating further into the motives of either man to initially undertake this search mission, Richardson used it as an opportunity to collect, experiment, and make observations on a broad range of natural history subjects. The expedition lasted only a year, yet Richardson collected enough material to fill a two-volume narrative with detailed accounts which fit in with the wider discussion of the HBC territories' governance. Richardson's Arctic science had economic and geopolitical implications, and as the next section shows, the differences between the way the two men prioritized their time in the Arctic also had significant implications for the reception of Rae's later report to the Admiralty which conveyed the first intelligence about the fate of the lost Franklin expedition.

CANNIBALISM AND THE QUESTION OF INUIT TESTIMONY

Prior to his expedition with Richardson between 1848 and 1849, Rae had spent time in the Arctic surveying Boothia Felix to determine if it was a peninsula. This expedition had already been suggested in 1840 and was supposed to have been under the command of Thomas Simpson. Simpson's untimely death paused the plans until they were renewed by the governor in chief of the HBC territories George Simpson in 1845. Rae's one and only narrative was based this expedition, published in 1850 as *Narrative of an Expedition to the Shores of the Arctic Sea, in*

1846 and 1847. The *Athenaeum* was not the only newspaper celebrating Rae's accomplishments. His narrative and his person were generally described as a perfect example of a modest, competent, and brave Arctic explorer. As one review noted, "To that gallant band is now to be added the name of John Rae; who with power of endurance combines excessive fortitude and coolness in the hour of danger. His high moral and physical qualities won the esteem and admiration of Sir John Richardson,—and the unpretending narrative now before us will tend to confirm the sentiment pre-existing in his favour."[33]

Comparing this to Richard King's thundering criticism of Rae just a handful of years later gives a good sense of the impact caused by Rae's report on the fate of the Franklin expedition. In 1855 King published a polemic book *The Franklin Expedition from First to Last,* wherein he stated that "I had all along associated Dr. Rae with the members of the medical profession who have distinguished themselves as travellers, such as Park, Oudenay, Richardson, McCormick, Daniel, Leichardt, and Kane; but I now find, and I rejoice in the discovery, that he is what he signs himself—as 'C.F.,' that is to say a Chief Factor, a *trader* in the service of the Hudson's Bay Company."[34] According to King, Rae had lost all credibility as an Arctic explorer and as a man of science because of his reliance on secondhand information from Inuit.

Franklin's expedition was last seen by Europeans in July 1845, and what happened after that has been clouded in mystery.[35] The specification "Europeans" is significant, as the trustworthiness of testimony from Indigenous peoples who reported sightings of Franklin and his crew became an issue of controversy when Rae was thrown into a large and very public debate over the content of his report and the value of his evidence. Rae did not discover the fate of the Franklin expedition during his search mission with Richardson; rather, he received the intelligence while surveying for the HBC. He was on his way from Boothia toward Repulse Bay when he met a group of Inuit from Pelly Bay. From them he obtained both relics belonging to the Franklin expedition and information about their deaths: Franklin's men had died with the last survivors resorting to cannibalism. Rae sent a short report based on this information to the British Admiralty, dated July 29, 1854. Rae also sent a letter to the George Simpson on September 4, 1854. Both Simpson and the British Admiralty proceeded to send his letters to the press, apparently without Rae's knowledge. They were published in full in multiple newspapers on October 23, 1854.[36]

The immediate response to Rae's report was mixed, but three key points can be drawn: First, the area where Franklin's men had been seen was where

FIGURE 3.1. Images of the relics belonging to the crew on board the HMS *Erebus* and *Terror* were popular motifs in the illustrated periodical press and books about Arctic exploration. *Source*: W. H. Davenport Adams, *The Arctic World: Its Plants, Animals, and Natural Phenomena*, 1876, 235.

King had proposed to search, but his suggestion had been rejected by the government. Second, the Admiralty was strongly criticized for not doing enough to save Franklin and his men. As the *Daily News* noted, Rae's discoveries "render more heavy than ever the moral responsibility and the professional guilt of those whose immediate duty it was to rescue a body of gallant men long within reach of help, but now lost to us for ever."[37] Third, the extent to which Rae's evidence was sufficient to determine the fate of the Franklin expedition was questioned. That the expedition had resorted to cannibalism was not the news Lady Jane Franklin wanted to hear, and Rae was subsequently condemned by many prominent British figures, including Charles Dickens (1812–1870). Rae did not find Franklin himself but based his conclusion about the expedition's fate on what he was told by Inuit. To what extent this was sufficient evidence became a key point of controversy and had consequences for Rae's social and scientific standing. The early responses to Rae's report are telling, not only of the prevailing racist views of Arctic Indigenous peoples, but also of the situational acceptance of secondhand observations in the making of Arctic knowledge. When, how, and why could explorers use information gathered by others?

Consider Dickens's criticisms of Rae. It is well known that Dickens's work was full of racist and anti-Semitic caricatures, embedded within an overarching belief in the moral superiority of the British and righteousness of the Empire. Dickens's two-part essay, "The Lost Arctic Voyagers," for his own weekly magazine, *Household Words* (est. 1850), was a tour de force of such stereotypes.[38] Dickens's criticisms of Rae, echoing that of Jane Franklin, had as their premise that Inuit were amoral and untrustworthy and that Rae was wrong to rely on their testimony. Dickens argued that Rae's testimony was founded on the mistaken belief that his own encounters with Inuit could give indications as to how they would behave if they were in a position of power: "It is impossible to form an estimate of the character of any race of savages, from their deferential behaviour to the white man while he is strong. . . . We believe every savage to be in his heart covetous, treacherous, and cruel; and we have yet to learn what knowledge the white man—lost, houseless, shipless, apparently forgotten by his race, plainly famine-stricken, weak, frozen, helpless, and dying—has of the gentleness of Esquimaux nature."[39] Was it not more plausible, asked Dickens, that it was a group of Inuit who had committed cannibalism when they found Franklin and his men? Perhaps they had even been murdered by Inuit.

While Dickens did not explicitly state that Rae was not as trustworthy as other British explorers, this was the implication of his essay, positioning Rae against Franklin and linking to broader questions of the status of scientific researchers and fieldwork. Whereas part one of "The Lost Arctic Voyagers" focused on the content of Rae's report to the Admiralty, part two consisted of an anthology of previous situations where British sailors had been lost, without food or water, and had not resorted to cannibalism. By juxtaposing Rae's testimony against that of other accounts of British men who had been in similar situations, Dickens made it a question of "the nature of men"—and whom one should trust, Rae or Franklin. If Rae was correct, then Franklin was amoral, worse than all other British men before him. Therefore, Dickens emphasized, Rae had to be wrong. Dickens himself had never traveled to the Arctic. To make his case against Rae, a well-known seasoned Arctic explorer and HBC employee, Dickens drew on the often-used rhetorical strategy of summarizing, or anthologizing, what other firsthand observers had experienced and reported, a well-established technique for establishing authority, one Rae utilized as well.

Throughout the first half of the nineteenth century, so-called armchair scholars relied on the knowledge that others had collected in the field.[40] Those working in the field were often of a lower social status than the gentlemen-scientists who made use of their collected data and specimens. Richardson is a useful point of comparison. He was an established and respected naturalist and was knighted for his services. While he was a Scottish surgeon and Arctic explorer like Rae, he was part of a tradition of explorers who had considerable social status. They traveled in a style much different from that developed by the HBC, as well as those associated with the KGH. Arctic expeditions were always dangerous and arduous, but facilities on board British naval vessels in the Arctic mirrored the gentlemanly status of its officers. Rae, however, made full use of Inuit methods during exploration; for example, he became skilled at walking in snowshoes.[41] He was rugged and distinctively non-gentlemanly. Up to a certain point, this approach was considered an asset, but it could also be used against Rae to discredit his arguments. In 1850 Jane Franklin wrote to Rae thanking him for his efforts, noting that "henceforth it is not as an officer of superior rank" in the HBC that he would be "solely regarded, but as one of our most distinguished navigators & travellers whose value will go down to posterity."[42] Jane Franklin's estimate of Rae changed significantly following his report to the Admiralty in 1854. In November of that year, she wrote to the Admiralty to warn them against letting the HBC organize an expedition to verify Rae's account, as the "Hudson's Bay Officers are not capable, as I learn from Dr. Rae, of making observations for latitude & longitude, and consequently are liable to go wrong & unable to direct their steps with certainty & confidence, if they come upon new ground." Furthermore, as this mission "is of a peculiar and almost sacred character," it would not be advisable to hire people without the "enlightened zeal" required of British officers.[43]

It is interesting to observe, how Jane Franklin wrote to her solicitor in May 1855 describing when she "felt & knew" that she was a widow. It was not in early 1854, when the Admiralty informed her it would remove the officers of the *Terror* and *Erebus* from the list of living officers, but it was the "unexpected intelligence" in "Dr. Rae's tragic letter" and the "relic he had obtained from the Esquimaux" that convinced her that her husband "had ceased to live." Though she privately took Rae's information as evidence that the men on board the *Erebus* and *Terror* were dead, her semiprivate and public responses were different. In these she argued that Rae's information was "not proofs of any theory," referring to the manner in which the crew had died. As she further noted, the "hails of horror" in Rae's report "ought never to have been published or even worded."[44] The tension in Jane Franklin's letter to her solicitor—between on the one hand accepting that Rae had discovered evidence about the expedition and on the other rejecting his "theory" about their fate—was reflected in the general reception of Rae's report in Britain. Indeed, as the historian Janice Cavell has observed, Jane Franklin utilized the information brought back by Rae to organize further expeditions.[45]

Another person who sought to use Rae's report while simultaneously delegitimizing him was Richard King. There is no doubt that King felt vindicated by Rae's report, as it showed he had been correct that the search missions should focus along the Back River and west of Boothia. The big issue was the financial reward for rescuing Franklin or determining the fate of his expedition. Neither Rae nor King was independently wealthy, and the reward was substantial. If Rae's findings were deemed sufficient, he would receive the reward. Yet, as King had proposed an expedition to search in that area as early as 1847, to what extent did he also deserve the reward? King thought he did. In 1855 he made his case in *The Franklin Expedition from First to Last*, which included a compilation of his correspondences with the Admiralty, as well as letters published in the periodical press. He extended his criticism to the Admiralty, and went so far as to include a "statistical form" of those on the board of the Admiralty who had been involved in the search for the Franklin expedition "in order to mark the exact amount of guilt which lies at each man's door."[46] In spite of his attacks on the Admiralty, King was short-listed as a recipient for the reward, which he did not receive.[47] In his strongly worded book, King argued that if he had been in charge of a search mission, he would not have relied on the words of Inuit but continued to investigate further:

> That he should have stood on the shore of Castor and Pollux River, his right eye directed to Point Ogle and his left eye to Montreal Island, knowing that the fate of The Franklin Expedition was to be read there, and instead of directing his steps to the tragedy before him, that he should have turned his back upon these painfully interesting lands, and have proceeded upon his paltry discovery, was utterly worthless, is a problem I will not pretend to solve. I was able to solve the problem of three centuries, the North-West Passage, in 1845, although it was not proved until 1854. I was able to point out the Death-spot of The Franklin Expedition in 1845, although it was not discovered until 1854; but Dr. Rae is a problem I cannot solve. He is a *conundrum* I *give up*.[48]

While King attempted to convince the public and the Admiralty that it was he who deserved the reward, at the same time he used Rae's report to justify that claim, while concurrently arguing that Rae's evidence was insufficient because he had relied only on secondhand information, thereby missing his opportunity to secure better and more substantial evidence.

For a further point of comparison on how the use of secondhand information was shaped by the broader context of the explorer's persona and his textual strategies, the American surgeon and explorer Elisha Kent Kane (1820–1857) is a useful study. Kane acted as surgeon to the first Grinnell expedition in 1850, before he was commissioned as commander of the second Grinnell expedition between 1853 and 1855. Both expeditions were privately funded by the American philanthropist Henry Grinnell (1799–1874) at the encouragement of Jane Franklin. Kane believed that the area around the Geographic North Pole was covered by a navigable Open Polar Sea, which could be accessed through Smith Sound. He thought it likely that the *Erebus* and *Terror* were drifting in this open water, unable to return south. This was highly hypothetical. Kane consulted the isothermal charts and theories of Humboldt and Dove, as had Richardson and the American oceanographer Matthew Fontaine Maury (1806–1873), in order to predict the lived experience of the Arctic.[49] While Richardson's focus was on accounting for the distribution of flora and fauna in relation to the potential for agricultural development, Kane looked to the temperature lines to predict the existence of the Open Polar Sea. While Kane did not find the crew of the *Erebus* and *Terror*, he claimed that his expedition had observationally proved the existence of the Open Polar Sea. However, he never saw this open basin of water himself. His "remarkable geographical results," as Jane Franklin's niece Sophia Cracroft (1816–1892) described them in a letter to Richardson in 1856, were based on the observations of William Morton, a steward, and Suersaq (ca. 1834–1889), a Greenlandic Inuk translator and dogsled driver.

Kane's claims to discovery based on secondhand information were not initially met with the same scrutiny as those of Rae. While certainly reflective of the contemporary negative attitudes toward Arctic Indigenous peoples, it is also suggestive of the fraught social context for the performance of Arctic expertise. Kane had had no Arctic experience prior to his participation in the first Grinnell expedition, but he quickly succeeded in framing himself as an expert on Arctic matters. After his first voyage, Kane embarked on a publicity tour to raise interest in his forthcoming narrative, as well as to gather support for his future venture. As Grinnell wrote to Jane Franklin in 1852, "Dr Kane's lectures have done much good, in giving confidence in the Public minds that Sir John may yet be living."[50] In addition to having the backing of key figures in the American Geographical and Statistical Society and the US Naval Observatory, Kane also had the support of key American political and cultural figures. During the period of Kane's Arctic ventures, his father, John K. Kane (1795–1858), was a judge of US District

FIGURE 3.2. Portrait of Suersaq as a young man, drawn by Elisha Kent Kane. *Source*: Elisha Kent Kane, *Arctic Explorations*, 1:24. Courtesy of the Scott Polar Research Institute.

Court for the Eastern District of Pennsylvania and a prominent figure within the American Philosophical Society. Kane drew on these connections when he applied to participate in the first Grinnell expedition. Grinnell was similarly connected with many influential figures. On several occasions he described in letters to Jane Franklin how he was lobbying politicians, including the president of the United States, for support.[51] This serves to show a key difference in the ventures of Rae and Kane which is likely to have influenced perceptions of their use of secondhand evidence. Jane Franklin and her network were *primus motors* behind the both Grinnell expeditions, and the stakes were high. As was the case for past privately funded ventures, such as that of Booth and Ross, the responsibility for the outcome of the expeditions rested with the organizers as well as the explorers. This became particularly problematic for Jane Franklin following the publication of Kane's second narrative, which I will return to in the next section.

Given that Jane Franklin was continuously working to raise backing for further search missions, she would have been unlikely to publicly criticize Kane for doing something very similar to Rae's actions. Just as Rae had decided the safest course was to immediately return with his information rather than personally validating it, Kane had also not wanted to risk journeying to verify Morton and Suersaq's observations and venture into the supposed Open Polar Sea to find the *Erebus* and *Terror*. In announcing his results, Kane made it clear that he had the utmost trust in the abilities of Morton and Suersaq to faithfully report what they had seen. Both Kane and Rae also combined their secondhand information with material evidence. In Kane's case, the hypothesis on the Open Polar Sea inferred a milder climate with increased wildlife toward the North Pole, which Kane believed was evidenced in increased hunting dividends during Morton and Suersaq's journey. Rae purchased several relics belonging to the lost Franklin expedition, and he cross-compared what he was told about the expedition from different groups of Inuit. In this way both men appeared to have made use of very similar repertoires for gaining knowledge in and about the Arctic. This leaves another point of comparison, namely the identities of their informants. Although Kane named Suersaq in his description of their surveying work, it was Morton who was given credit for seeing the Open Polar Sea. As Kane wrote in his narrative, at Cape Constitution "Mr. Morton saw no ice."[52]

As the criticisms of Dickens and Jane Franklin show, the fact that Rae had relied on Indigenous people for information played a key role in his delegitimization. The historiography on Rae is full of wildly contrasting accounts of his views on Indigenous peoples. For example, Russell Potter argued that while Rae's views of Inuit was not always positive, through "long and direct experience" and "unlike naval explorers, who tended to regard them as a dirty, uncivilized, and unreliable race, Rae came to respect and admire them, and counted many among them as his personal friends." Rae's biographer Ken McGoogan further argued that during the controversy, and "at considerable cost to himself, Rae stoutly defends the Inuit." By contrast, Janice Cavell described Rae's attitude toward Inuit as much more cynical, noting that he "believed the Inuit not so much because he considered them honest as because he considered himself well able to see through them when they lied: surely an exaggerated claim from a man who, for all his long northern experience, had spent relatively little time with these people and did not understand their language."[53]

Rae's *Narrative of an Expedition* as well as his personal correspondence certainly contain several episodes to support Cavell's interpretation. For example, regarding Inuit intelligence, Rae described his interpreter Ivitchuk[54] from Repulse Bay as "too stupid" to think about informing him about canine preferences for different types of seal fat until it was too late to save them.[55] Regarding truthfulness, Rae described how a man named Ak-kee-ou-lik[56] told a lie that Rae "did not believe at the time, and I afterwards found out that it was false."[57] While it is important to acknowledge that Rae was still acting and writing from the standpoint of Eurocentric racial stereotypes—as both Cavell and Potter point out and his private letters substantiate—it is also significant that Rae differed from many other British explorers on these issues, in particular because it affected both his Arctic science and the later controversy over the fate of the Franklin expedition. As Potter further argued, Rae "was accused of accepting second-hand evidence from a savage people, a race with a 'domesticity of blood and blubber' (in Dickens's words)."[58]

While Rae, Kane, King, and Richardson should not be considered fully contrasting figures, a comparison of the differences and similarities in their styles of exploring, writing, and social status can show how a wide range of factors influenced how the Arctic was represented. For example, Rae's *Narrative of an Expedition* and Richardson's *Arctic Searching Expedition* were stylistically very similar: both were mostly void of the types of rhetorical strategies utilized by other writers to generate interest, something which, as the next section shows, both the American Kane and the Dane Petersen did. As one reviewer in the *Spectator* noted about *Arctic Searching Expedition*, "It is rather a book of important scientific facts and observations than of travel or adventure." Rae's expeditions were widely commented upon in the periodical press, but his *Narrative of an Expedition* was not widely reviewed there. Advertisements and reviews appeared primarily in the *Athenaeum*, the *Quarterly Review*, and the *Examiner*.[59] It's possible the publication of his narrative was simply overshadowed by the ever-growing interest in the lost Franklin expedition. Richardson's narrative was widely noticed, with long reviews appearing in the *Dublin University Magazine*, the *Examiner* (jointly with a review of *Narrative of an Expedition*), the *North British Review*, the *Athenaeum*, and shorter mentions in several other publications. Richardson's narrative dealt directly with the Franklin question, and this could be a significant factor in why it received more attention than Rae's narrative of his previous expedition. As the *Quarterly Review* noted, "It is curiously illustrative of the interest excited by this expedition that Richardson received numerous advances from volunteers desirous of joining him."[60] We might infer from this that the efforts to discover the lost expedition both supported search missions as well as an industry surrounding them, including their narratives.

In the years that followed, Jane Franklin found out through Grinnell that he was hoping to publish his journal, which she noted in her letter book "was bad news for us!" But writing did not come easy to Rae, and he did not complete the project. He had had difficulties in the publication process of his travel narrative from 1850 as well, and mentioned them in letters to Richardson in 1849. His manuscript had been sent to the Royal Navy officer and hydrographer Francis Beaufort (1774–1854) and the HBC secretary Archibald Barclay, but "Mr Barcley has not honoured me with a line although I have written him by every opportunity." Perhaps, he wondered, Barclay had "thought it not worth publishing, and most probably it is not." Nevertheless, even though Rae had little confidence in his narrative and noted that Barclay "may be of the same opinion as myself that it would make by a sorry book," he emphasized that the book itself was an honest representation of his expedition, with no "falsehood nor exaggeration in it but rather the contrary."[61]

Yet constructing a trustworthy Arctic account was more complicated than simply stating that it was void of exaggeration and falsehood. Acceptance of observations and theories related to Arctic phenomena was fundamentally tied up with perceptions of the author. Rae's report to the Admiralty had long-lasting consequences. After several years' delay, he was eventually given the reward for ascertaining the lost expedition's fate. However, Jane Franklin lobbied to delay paying him, and at the same time the HBC suspended Rae's pay. Rae had to send several letters to the Admiralty and the company before the money finally was released. As he noted in a letter to Richardson in 1855, "I do not think their Lordships or the Company have behaved very fairly in my case." Aside from the financial issue, there were other consequences for Rae. When King argued that Rae had shown himself to be nothing more than a fur trader, he delegitimized him as an explorer and consequently his discoveries, both geographical and scientific. This attack on Rae was not limited to King or Dickens. Ken McGoogan has shown that the naval hydrographer John Washington attributed the charting of Victoria Island to Richard Collinson (1811–1883), even though Rae had charted it two years before. Washington further argued against giving Rae the reward for finding Franklin. Rae also experienced controversy at a meeting of the Geographical Society in 1856, where his geographical discoveries were questioned.[62] Rae's decision to convey and continuously defend the information he had been given by Inuit as certain proof of the Franklin expedition's fate seriously harmed his reputation. The implication was that Rae had "gone native" and could no longer be trusted as a British gentleman. While he was able to retire on the reward money, unlike the majority of the British leaders of Arctic expeditions, Rae was never knighted and his past geographical discoveries were downplayed—though in his later career they played a key role in the organization of the Canadian-British contribution to the International Polar Year. Rae's rugged persona, his abilities to travel and survive in the Arctic and his decision to trust Inuit testimony were used to discredit him as an Arctic explorer. Other explorers who wished to seize an opportunity for employment as part of an Arctic expedition and a chance to be part of discovering the Northwest Passage could simply discard Rae's evidence.

A DANISH OPPORTUNIST

Even after Rae received the reward for ascertaining the fate of the lost Franklin expedition, Jane Franklin did not cease her campaigns for finding the now-presumed-deceased men. As part of her efforts, she had secured the support of high-standing scientific men, such as the president of the Royal Society, Roderick Murchison, to assist with lobbying for further resources for expeditions. Murchison and thirty-five other prominent British men, and eighteen officers from the Royal Navy who had been employed in the search for Franklin, signed their name to the letter.[63] It urged the British government to send out an expedition to further examine Rae's claims. Such a project, it argued, would be of little risk, as Rae's report directed them to a limited geographical area. This is another example of the conflict caused by Rae's report, between trusting and using to their advantage the fact that Rae had discovered *where* Franklin and his crew could be located on the one hand, and disputing *what* had happened to the expedition on the other. As the letter stated, a search expedition could "satisfy the honour" of Britain, and "clear up a mystery which has excited the sympathy of the civilized world." The lost expedition generated international interest and in turn international collaboration and financial assistance for search missions. In 1849 the English politician Sir Robert Inglis described to the British House of Commons how assistance from America and Russia meant that "the three greatest empires in the world, had co-operated heartily, not in schemes for their own aggrandizement, but for the relief of suffering humanity." Jane Franklin worked to foster this international interest resulted in ventures such as those funded by Grinnell. Yet people like McClintock expressed concern that as a consequence of this interest, a foreign nation might discover the fate of the lost men. As he wrote in his narrative, the issue was "a *great national duty*."[64]

There was much geopolitical unrest in the middle of the nineteenth century, but also increasing international collaboration. In particular, significant advances in geomagnetism were made through international collaboration on terrestrial magnetic observations. The enthusiasm behind this is well captured in its designator, the magnetic crusade. Geomagnetism was of great importance to navigation, as establishing a theory for the Earth's magnetic field could explain the long-observed variations in the compass. Improvements from these ventures included the improvement and standardization of instruments such as the magnetic compass and dip circle. For example, aside from geography and trade, the central focus of the Franklin expedition had been geomagnetism. Observations were made and shared between multiple countries—including Britain, France, Prussia, and the United States—and carried out both in the Arctic and Antarctic. Throughout the nineteenth century there was a tension between nationalism and attempts at international scientific partnerships in the Arctic, as throughout the globe. However, as Marc Rothenberg has argued, "the Magnetic crusade was . . . more of a limited international cooperative venture rather than a true collaboration."[65] Such international collaborations were not so much between governments, as in the years leading up to the International Polar Year examined in the next chapter, but between individual figures and expeditions.

In spite of efforts such as that of Murchison and the cosigners of his letter, the British Admiralty was not interested in spending more resources, financial or human, on the subject. Jane Franklin therefore organized her own expedition partially funded through a public appeal, led by Captain Francis Leopold McClintock (1819–1907) on the steam yacht *Fox*. McClintock is a good example of how her efforts created new opportunities for officers seeking employment. In 1855 he wrote to James Clark Ross (now a senior Arctic and Antarctic explorer) that although "Arctic work is now over," he was "hoping for a ship" where his "ice-experience may be useful." Perhaps, McClintock wondered, Ross would "assist in launching me off once more." A few weeks later, McClintock followed up with another letter to Ross, stating that he had "been working at the Admiralty for Employment, & have some hope, although slight, for success." McClintock was eager to return to the Arctic, and he brought this enthusiasm to Jane Franklin's *Fox* project. It was McClintock who inspected the *Fox* and convinced her that it was a good purchase. She officially offered him command of the expedition in April 1857, which he accepted immediately.[66]

In addition to the support Jane Franklin had secured from high-standing figures in scientific and political circles, there was also still a significant international public interest in discovering more about the lost expedition's fate, which was reflected in the crew of the *Fox*, one of whom was the Danish translator and experienced Arctic explorer Johan Carl Christian Petersen. Following the expedition, both Petersen and McClintock published accounts of their experiences, with McClintock's *The Voyage of the "Fox" in the Arctic Seas*, published in 1859, and Petersen's *Den Sidste Franklin Expedition med Fox* the year after.[67] Petersen was born in Copenhagen into a family of little means. According to Niels Aage Jensen, Graah's expedition in Greenland (see Chapter 1) was widely discussed in the 1830s, and would likely have strengthened Petersen's desire to travel to Greenland. He moved to Qeqertarsuaq (previously Godhavn) in May 1833. After a hand injury left him unable to continue working as a cooper, Petersen secured a position relating to trade and hunting with the KGH.[68] Petersen was not a scientific man. Aside from his basic education in military school, his training was practical. But through his time with the KGH he became well versed in multiple languages and was especially known for his Greenlandic language skills and strong knowledge of Greenlandic culture. He was also skilled at traveling by dogsled. This was reflected in his narrative of the McClintock expedition, which primarily focused on ethnography, linguistics, and geography, in addition to its search mission.

Throughout his narrative, Petersen included detailed accounts of his encounters with Inuit. This served to illustrate his intimate knowledge of the language and culture but also reveals the friendships he had formed during his previous expeditions. For example, while in the northwest coast of Greenland the expedition encountered a group of Inuit men whom Petersen knew from an expedition with Kane. They all recognized each other, but he noticed that Suersaq was missing. Petersen related how Kane had threatened Suersaq multiple times, as Kane had claimed he had "the right to have him shot for his disobedience." Suersaq had been able to escape (Kane used a different term, *desertion*), and Petersen was happy to learn that he had since married. McClintock and his men wanted to purchase sled dogs from Inuit, but Petersen noted, "These people, who have shown themselves to be so helpful and respectable towards us during the unlucky voyage we did from Advance to possibly escape down to Upernavik, had suffered much since that time," and there were no dogs available. McClintock noted in a private letter that he had been told that Suersaq wanted to relocate to the Danish settlements in southern Greenland, but "he has neither dogs nor Kayak."[69]

Petersen's descriptions of encounters with Indigenous peoples furnished important evidentiary resources for ethnographers. He also collected clothes and other objects, which were later exhibited in Copenhagen. Following the *Fox* expedition, Petersen received several medals and honors, including the British Arctic Medal, the Swedish Polar Star, and the Danish Order of the Dannebrog. His sled flag was displayed at the Ethnographic Museum in Copenhagen. As was reported widely in the Danish press, Jane Franklin presented Petersen with a pocket-watch with an engraving of the *Fox* as an acknowledgment of his service during the expedition. Petersen also gave several lectures about the voyage, including at the Group for Industry in Copenhagen (Kjøbenhavns Industriforening).[70] His narrative was completed with the assistance of Frederik Wøldike (1832–1883), who also published his book.

Petersen's language skills allowed him to work as a translator on several Arctic expeditions. Between 1850 and 1851 Petersen participated in Captain William Penny's search for the lost Franklin expedition on board the ships *Lady Franklin* and *Sophie*. The expedition was delayed by ice around Upernavik, and a brief description of this delay nicely illustrates the Franklin fever of the time. While delayed, they were met by the expedition in the HMS *Resolute* and *Assistance* and the steamers HMS *Pioneer* and *Intrepid*, led by Captain Horatio Thomas Austin. A few days later yet another expedition arrived, the *Prince Albert* with Captain Charles Codrington Forsyth, which was also financed by Jane Franklin. John Ross's last expedition to the Arctic on board the *Felix*—named after his patron—also arrived. Outside Lancaster Sound they met the first Grinnell expedition's USS *Advance* and *Rescue*, led by Lieutenant Edwin de Haven. Inside the bay they met the supply ship *North Star* under Commander James Saunders[71] which had left to provide assistance to James Ross who had also been looking for Franklin. In total there were eleven ships in the Barrow Strait by the mouth of the Wellington Channel outside Beechey Island.[72] The lost Franklin expedition provided plenty of opportunities for Arctic explorations, funded by governments of several countries and various types of private patrons.

Following Penny's expedition, Petersen was asked to participate in the second Grinnell expedition. In a letter to the American secretary of the Royal Navy, later published in his travel narrative *Arctic Explorations*, Kane described Petersen in flattering terms. He wrote that he had "engaged the valuable service" of Petersen, so that if they "should meet the Esquimaux north of Cape Alexander, he will be essential to our party." Petersen was part of the crew from the second Grinnell expedition who were unhappy with Kane's leadership, as well as with the prospect of spending a second winter in Smith Sound. The crew therefore left (Kane again used that other word, *desertion*) for an unsuccessful search of Upernavik. Petersen worked to keep the crew alive during the second winter. Reflecting on this, McClintock noted that "the Danes say that Kane's crew told them that but for Petersen they would all have starved!" McClintock also wrote to Jane Franklin that in Greenland "they regard him [Petersen] as a sort of 'flying dutchman'! & listen with open mouths to his wonderful stories. There is but one opinion in Greenland & that is that he saved Kane's expeditions. I like him very much."[73] After his return, Petersen and his family visited Denmark. Immediately upon his arrival, Petersen was asked by the chamberlain and naval officer Carl Ludvig Christian Irminger (1802–1888), royal adjutant to King Frederik VII, if he would participate in McClintock's expedition at the request of Murchison. Murchison was, of course, working with Jane Franklin to build support for the search missions. Irminger had a keen interest in Arctic research, and was cofounder of the Royal Danish Geographical Society (Kongelige Danske Geografiske Selskab). He published several books on ocean currents and other geographical and hydrographical subjects, and the Irminger Sea and Irminger Current are named after him. As an "Honorary Corresponding Member" of the Royal Geographical Society of London, Irminger was well known to Murchison.[74] Petersen is in this way an interesting example of how the lost Franklin expedition fostered a high level of international collaboration in the Arctic.

In spite of the fate of the second Grinnell expedition, Petersen agreed to participate in McClintock's expedition. The *Fox* was already waiting in Aberdeen, and Petersen left Copenhagen by train soon after. In London he had lunch with Murchison, and was gifted a map of the Arctic regions by John Washington (the same man with whom Rae had a dispute over the designators in his map). The expedition left Aberdeen on June 30, 1857 and returned to London on September 21, 1859. The *Fox* had previously been used only for leisure travel to Norway. Jane Franklin had purchased it for £2,000 pounds and had it reinforced in Aberdeen. All in all, McClintock estimated that they would require a total of £10,000 to purchase, reinforce, and outfit the *Fox* with provisions, as well as to pay the crew for two years. They had provisions for twenty-eight months. The expedition began with twenty-five people; in Greenland they added two Inuk men named Anton Christian and Samuel Emanuel,[75] while one of the expedition members returned to Denmark because of illness. The expedition had been provided with a letter from the directors of the KGH to Inspector of the North Greenland Christian Olrik (1815–1870), and he helped them obtain ten sled dogs.

According to McClintock, Olrik was particularly impressed by the "good conduct of the men . . . being so different to what is sometimes experienced in the arrival of English, Danish and American ships" and had a large quantity of beer brewed for the crew.[76] On August 6 they reached Upernavik, where Petersen had lived for twelve years and where they added fourteen dogs and a reserve of seal meat. They arrived in Melville Bay on August 12, but were soon caught by ice after until April 26. During these 242 days, they drifted 1,194 geographical miles through the pack ice. Reaching Beechey Island on August 11, they continued to Bellot Strait. They spent the second winter around Point Kennedy and in early spring divided into three overland search parties. In early March McClintock's party met a group of Inuit who confirmed Rae's information and sold them items from the Franklin expedition. They met several other groups of Inuit, who provided them with more information and relics. By following these reports, they discovered a skeleton in uniform on May 24. Soon after, one of the overland parties discovered a key piece of evidence: surviving members of the Franklin expedition had left a letter that, among other things, gave the date of Franklin's death. This letter was the smoking gun that provided what Rae had been criticized for not finding, namely evidence based on something other than Inuit testimony. Rae reflected on this in 1877: "I need hardly repeat how this story was confirmed in all important particulars by the information and document brought home by the Fox Expedition in 1859."[77] The letter was undisputable proof that Franklin was dead but did not shed light on the issue of cannibalism.

Petersen's and McClintock's published narratives addressed Rae's report, but in very different ways. McClintock's was guarded and did not explicitly discuss the question of cannibalism, while Petersen used it to add drama to his narrative. For example, he recounted how in 1854, during the Kane expedition, he had met Rae while examining the area around Boothia. Petersen described Rae's discovery of the Franklin expedition's fate in dramatic terms, noting that "the Eskimoes assumed, that they had starved to death after they in vain had tried to save their life on each others flesh." Similarly, a review of *Den Sidste Franklin-Expedition* noted that "the Eskimoes reported that the white men had passed away while travelling to Fish River, and this testimony was confirmed by human remains that they found along their way." Petersen further described

H.M.S.*hips Erebus and Terror

28 of May 1847 } Wintered in the Ice in
Lat. 70°.5′ N. Long. 98°.23′ W

Having wintered in 1846—7 at Beechey Island
in Lat 74°. 43.28″ N. Long 91.39.15″ W After having
ascended Wellington Channel to Lat 77°. and returned
by the West side of Cornwallis Island.

Sir John Franklin commanding the Expedition.
All well

Party consisting of 2 Officers and 6 Men left the Ships on Monday 24th. May 1847

Commander.

WHOEVER finds this paper is requested to forward it to the Secretary of the Admiralty, London, *with a note of the time and place at which it was found:* or, if more convenient, to deliver it for that purpose to the British Consul at the nearest Port.

QUINCONQUE trouvera ce papier est prié d'y marquer le tems et lieu où il l'aura trouvé, et de le faire parvenir au plutot au Secrétaire de l'Amirauté Britannique à Londres.

CUALQUIERA que hallare este Papel, se le suplica de enviarlo al Secretarie del Almirantazgo, en Londrés, con una nota del tiempo y del lugar en donde se halló.

EEN ieder die dit Papier mogt vinden, wordt hiermede versogt, om het zelve, ten spoedigste, te willen zenden aan den Heer Minister van de Marine der Nederlanden in 's Gravenhage, of wel aan den Secretaris den Britsche Admiraliteit, te London, en daar by te voegen eene Nota, inhoudende de tyd en de plaats alwaar dit Papier is gevonden geworden.

FINDEREN af dette Papiir ombedes, naar Leilighed gives, at sende samme til Admiralitets - Secretairen i London, eller nærmeste Embedsmand i Danmark, Norge, eller Sverrig. Tiden og Stedet hvor dette er fundet önskes venskabeligt paategnet.

WER diesen Zettel findet, wird hierdurch ersucht denselben an den Secretair des Admiralitets in London einzusenden, mit gefälliger Angabe an welchen Ort und zu welcher Zeit er gefunden worden ist.

how Franklin's earlier 1819 expedition had suffered greatly, to the point that "his men's hunger had been pushed so far, that they thought they had to use this last, gruesome rescue tool—to feast on the meat of a friend; only by using force had Franklin ensured that there was only one victim." Cannibalism was not foreign to Arctic expeditions, and Janice Cavell has argued that "it must be remembered that survival cannibalism was not repudiated by nineteenth-century European mores, and that it was white men who actively spread the allegations."[78] However, as the British reactions to Rae's report illustrates, while it was not a complete surprise that survival cannibalism might happen in the Arctic (and elsewhere), it was deemed improper to speculate about it.

What was proper to include in a narrative and what was not are likely to have been of particular concern to McClintock because of the reactions to Kane's narrative from the second Grinnell expedition. Jane Franklin was initially pleased with the results of Kane's expedition, in part because the supposed discovery of the Open Polar Sea provided a boost for the theory that Franklin and his crew had made their way to this body of water and had traced the Northwest Passage. It had, however, been a miserable expedition. As Petersen put it, "This was a very unlucky voyage, long-lasting and without any results"[79] They were icebound on the coast of Greenland for two winters, they did not have enough fuel, the ship was not properly insulated, the crew suffered from an outbreak of scurvy, and several men died. Kane described these difficulties in vivid detail in his second narrative, which was problematic for Jane Franklin, as she was hoping to build support for further search missions. In fact, Kane's narrative was so explicit that it caused several people to retract their support for her proposed new private ventures. She likely felt that she could not criticize Kane publicly without alienating her American supporters, including Grinnell, but she expressed her concerns in private letters. In a letter to Grinnell, Sophia wrote that her aunt deeply regretted having encouraged others to read the narrative. "She feels it all the more," Sophia wrote, because letters included in the narrative represented her "as having urged Dr. Kane to undertake the expedition." Sophia and Jane Franklin believed the "destructive effect" of the narrative caused "fearful obstacles" for the organization of future Arctic expeditions, private or public.[80]

◀ FIGURE 3.3. The expedition under the command of Francis Leopold McClintock found a letter that gave the date of John Franklin's death, which was reprinted in multiple books in the years that followed. *Source*: Carl Petersen, *Den Sidste Franklin-Expedition*, 1860, plate 19.

Kane had been sickly his entire adult life and was dying by the time his second narrative was published. He traveled to Cuba in an attempt to improve his health, but passed away in Havana in February 1857. His passing caused more problems for Jane Franklin. Sophia wrote to Grinnell that Kane's death and the planned publication of a biography of him were likely to increase the "public prejudice" against Arctic expeditions, "because it is evidently the desire in America to make it appear that his own Arctic adventures & sufferings" had been the cause of their premature return to the United States. She wrote that Kane's descriptions of those sufferings "portrayed by himself with such startling & graphic minuteness in his book have done our cause more harm than anything which has ever been written or spoken."[81] This letter was written in April 1857, the same month that Jane Franklin purchased the *Fox*, which set sail in July that same year. As it was, McClintock was not her first choice to lead the expedition, but she had difficulties finding anyone to take on the project.

In a letter to Grinnell, Sophia revealed that they were considering asking Rae to lead the expedition: "After all we had said of Dr Rae, you may at first be shocked that any application could be thought of to him," but her aunt "cares of course much more that the thing should be done, than who should do it." In fact, Jane Franklin "knows no one more fitted for it than Rae"—aside from Collinson, Richardson, McClintock, Maguire, and Osborne. Sophia further argued that even if Rae would be undertaking the role for "his own interests sake & for ambition" and not with "disinterested devotion," they both believed "he could do the work better than any one else—& that he would have no objection to undertake it." That is, Rae could be hired for the job. He was due to arrive in New York, and Sophia asked Grinnell to seek him out on their behalf. A month later Sophia expressed her relief that this had not come about, noting that they "rejoice to think that he was not at N.York, [so that] no proposal would be made to him."[82] Jane Franklin had such difficulties finding a commander following the second Grinnell expedition that she almost offered the job to a man she had spent years criticizing. This is highly suggestive of the degree to which the identity of the Arctic explorer was constructed within the wider context of their ventures. Jane Franklin argued that these problems were directly linked to the reception of Kane's second narrative. This is an important point to consider regarding the differences between McClintock's and Petersen's narratives. One was exceedingly cautious in his descriptions, while the other appeared to have no such concerns about directly engaging with the explicit details of Arctic exploration, including Rae's report.

The content of Rae's report had been widely circulated in the Danish press, and appeared in full, or extracts, along with detailed commentary. What we see from these reviews is a different attitude to the issue of cannibalism—and to the trustworthiness of Rae's testimony. An article in the *Berlingske Tidende* focused on the dramatic elements of the report, noting that "several of the bodies were horribly mutilated." Similarly, *Dannevirke* wrote that "some of the dead bodies were mutilated and robbed by the survivors." While Dickens and King questioned the trustworthiness of Rae's report, the articles in these newspapers simply stated that this was information collected by Rae through conversation with Inuit, and that Rae could see no reason to assume that the Indigenous peoples had caused the death of the explorers. Likewise, the article in *Fyens Stiftstidende*, while emphasizing the drama of Arctic exploration and cannibalism, maintained that Rae's report showed that "some of the bodies were buried in the snow, likely those who died first, while the others were spread around the ice, and some of the last wore clear evidence that the unfortunate in their hunger had reached for the last resource in the efforts to sustain their lives on their dead comrades flesh." An article in *Sjællands-Posten* went further, speculating on whether Franklin himself had committed cannibalism. The article also thought that because Rae had returned with one of Franklin's medals, and there was an officer among the bodies found on the ice, "it is likely, that it is himself, who was one of the last victims."[83] This attitude to cannibalism is indicative of a difference between what Franklin represented in Britain and in Denmark: Franklin was not a Danish national hero, and therefore the idea that he and his crew had resorted to cannibalism was not an offense to the national self-perception.

It was difficult to determine with certainty, Petersen conceded, what had been the end result of Franklin's last expedition. Wild animals could also have inflicted the wounds on the dead. Yet if the last survivors had in fact resorted to cannibalism, "it would have been too difficult for them to again feel human and be able to return to the civilized world; it would have been particularly difficult to convince themselves to return to England, where they would fear seeing the brothers and friends of those people they had devoured, and perhaps even killed to satiate themselves on their flesh and blood." While McClintock did not explicitly address the issue of cannibalism, he emphasized at several points that Inuit testimony could not be trusted. The information one could gain from Inuit was "vague" as "indeed all Esquimaux accounts are naturally so." As such, McClintock argued, it was up to their "own exertions for bringing to light the mystery of their fate."[84]

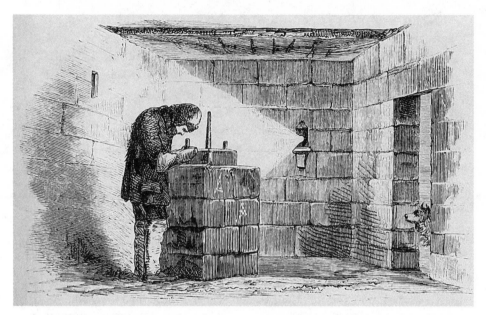

FIGURE 3.4. Illustration of an observatory constructed and used by the expedition under the command of Francis Leopold McClintock. *Source*: Francis Leopold McClintock, *The Voyage of the "Fox" in the Arctic Seas*, 1859, 206.

However, McClintock still relied fully on the assistance of Inuit to ascertain the fate of the Franklin expedition. How was this any different from what Rae had done? It is suggestive that McClintock put a lot of effort into making extensive scientific experiments and observations and the collection of specimens. The *Fox* voyage was a small expedition compared to other British expeditions, but it was still larger than Rae's overland expeditions. They were able to bring with them several scientific instruments, and he received training in their use and in preparing specimens by Sabine and Joseph Dalton Hooker (1817–1911). During the winter the expedition built magnetic observatories to record hourly observations. McClintock included these and other results in his narrative; a few tables appeared in Petersen's as well. This performance of scientific research was a central way in which explorers like McClintock could portray their broader observations and claims as trustworthy.

McClintock used Petersen's expertise to establish his argument that their interpretation of Inuit testimony was trustworthy. Dickens had harshly criticized Rae's interpreter during his 1854 expedition, considering him to not have been able to fully comprehend or convey important details. Interestingly, there is currently some debate on whether Petersen properly understood the dialect used in the central Arctic, yet this was not emphasized at the time.[85] In a letter to Jane Franklin, McClintock described his interactions with a group of Inuit who visited him on the *Fox* with information about the *Erebus* and *Terror*. He convinced them to draw a chart of Navy Board Inlet and noted that "according to their charts their usual route in winter lies partly over land." This had, McClintock wrote, "of course interested us greatly & this intercourse is very beneficial to Petersen" and the two men he employed in Greenland (their names were given as Anton and Samuel), as "their dialect differs slightly from this western one."[86] However, this problem and the extent to which they relied on Inuit knowledge were glossed over in both Petersen's and McClintock's narratives.

At the time of the *Fox* expedition, Petersen was an experienced Arctic explorer, and he was well known for his language skills and knowledge of Indigenous Arctic cultures. Petersen did not have the same need to differentiate himself from Rae. His wife Ida-Berthe was part Inuit, and he had adopted many of the Greenlandic traditions and ways of life. He introduced travel by dogsled to the British search expeditions, a skill he had acquired during his time in Greenland.[87] Whenever McClintock described subjects of ethnographic and linguistic interest throughout his narrative, he referred to Petersen in terms of excellence, experience, and intimate knowledge. In particular, he emphasized that it took skill to separate truth from falsehood in Inuit testimony, and that Petersen had the proper abilities. This praise was also repeated in the Danish press, which further highlighted his role in expanding collaborations between the Danish and British explorers:

> It is the British who take the main honor for exploring these areas. The Danish take the next spot after the British; Danish men eagerly participated in the early expeditions, the Danish colonies in Greenland have been of considerable importance for the later expeditions, and from there have the British received useful help in various directions.... [T]he man who has published this work, has surely not played a prominent role, has not been a leader of an Expedition, but he has in a subordinate role significantly contributed to facilitate the attainment of the planned objects.[88]

Who was considered a trustworthy observer was fundamentally linked to issues of the representation of the Arctic explorer. McClintock's guarded treatment of the subject of what had actually happened to Franklin and his men, his argument that Inuit testimony was vague and required correct interpretation, and the significant scientific results from the expedition appeared in contrast with Rae's. But Petersen noted that the whole issue, from the Northwest Passage to the search for Franklin, was founded on the British sense of national pride:

> Had England's interest in this question now merely had its foundation in the desire for commercial advantages, the Northwest Passage would probably never have been found, and there would hardly have been made even one additional attempt at finding it; but the question had in a sense become a point of honor for Britain and the British with their sharp minds do not like riddles which they could not solve; they continue to try and try—until they usually finally solve the riddle... . Finally they could no longer resist the old John Barrow's strong requests, which were supported by many other weighty voices as well as by public opinion, and the Franklin Expedition was fitted to depart in May 1845.[89]

The Northwest Passage was Barrow's life project, and, according to Petersen, it had been easy for Barrow to convince the government to fund further expeditions. The entire situation, he argued, was a very British problem: the government and the public could not bear the idea that they were not the discoverers of the passage. As Denmark was not financially able to support exploratory missions, Petersen was just happy to use the opportunity to get employment, as well as a chance at fame. Yet, he noted, perhaps it was best if the full details of the lost Franklin expedition remained hidden forever to "cover the misery that accompanied Franklin's brave men until the end, who had departed with *Erebus* and *Terror,* probably without thinking that the fate that awaited them was almost given in advance by the ominous names of the ships, which means darkness and horror."[90]

THE NEW CHALLENGES AND CONVENTIONS FOR SCIENTIFIC PRACTICE IN THE ARCTIC

The lost Franklin expedition changed the previous conventions of Arctic explorations. John Barrow had pushed for one last expedition in search of the Northwest Passage, and it ended terribly. While the historian Russell Potter has argued that the Franklin expedition caused a paradigmatic shift "as the nation's patriotic feelings seem to have been fueled less by the sublimity of sacrifice than by a sense of loss and mourning," the drivers behind the search missions and their representations were also shaped by the desire to undertake more expeditions to the Arctic.[91] The lost Franklin expedition afforded new opportunities for hopeful Arctic explorers, but because they were carried out under the banner of a great tragedy, the framing and configuration of the scientific discoveries within the British context had to be amended. This was not the case in the Danish or American contexts. It is also evident that while the primary objective of voyages was to determine the fate of the lost expedition, science remained an important focus. Scientific research was an important way for explorers to establish cultural and scientific authority in the period leading up to the disappearance of Franklin, and this function was further solidified in Britain after Rae's report. The opportunistic use or repurposing of science gives indications as to which tropes were considered most effective for establishing scientific credibility. In this way, the lost Franklin expedition simultaneously generated new opportunities for Arctic explorations, and challenged the conventions for the representation of the Arctic explorer and science.

At the same time, as an anonymous author in the *North Devon Journal* lamented, Arctic explorations were expensive, and it was not obvious to everyone that the scientific knowledge gained was worth the cost: "In no quarter of the globe has the spirit of geographical discovery made nobler efforts, displayed a more heroic endurance, or exhibited higher qualities of mind, that [*sic*] in encountering the difficulties and dangers of the Arctic Regions. Science has, no doubt, obtained some valuable accessions to her stores from those hyperborean sources; but they have been acquired by a large expenditure of money, and of—what is far more precious—human life and suffering."[92] The article was published following the disappearance of Franklin's last expedition. There had been other tragedies in the Arctic before this, but this was the first time an entire expedition had been completely lost. It had been Franklin's third expedition and if such a seasoned explorer and his crew could disappear, would future missions be worth the investment? Any answer to this question depended in part on how the Arctic was perceived as a field site for scientific observation. Participants in Arctic explorations were always motivated in ways varied and complicated in their relationship to official instructions. Although the tension between private and publicly funded explorations was not unique to the search missions, explorers faced the additional challenge of maintaining the perception that finding Franklin was in fact their primary objective, while balancing this with the established conventions for Arctic explorations. The (self-)portrayal of the Arctic explorer and his activities had a significant impact on the trustworthiness of their claims. The difference between Rae's and Richardson's prioritizations in their year together shows the discord between the stated aim, finding Franklin, and the produced results, the advancement of science with significant economic and geopolitical implications. While the British Admiralty had not stated any explicit scientific goals for the expedition, Richardson's findings were extensive and were linked to the concurrent debates over the renewal of the HBC charter. Whereas the company had claimed that their territories were unsuitable for settlement and had value only to the fur trade, Richardson's findings added to the arguments that there was in fact the possibility of extracting both food and mineral resources from the seemingly infertile land. By contrast, Rae's interests were geographic, and he spent the year surveying the coastline.

Richardson maintained the persona of the gentlemanly Arctic explorer even when the format for his venture broke with the typical blueprint. Richardson and Rae are an interesting example, as their activities before, during, and after their joint search of Franklin shows how the construction of the persona of the Arctic explorer as a gentlemanly observer of Arctic phenomena was central to the later outrage over Rae's claims. The reluctance to accept Rae's report to the Admiralty revealed a tension between how would-be explorers used Rae's findings to justify their proposed future expeditions, while simultaneously rejecting that Rae had been right to trust the testimony of his Indigenous informants.

Rae challenged the conventions of British Arctic explorations not only in his methods for traveling in the Arctic, but also because he openly prioritized Indigenous knowledge. The situation was different in Denmark. As there were no funds available to organize large Danish explorations, Petersen used the lost Franklin expedition as an opportunity for employment. In comparison with McClintock's published account, Petersen's narrative revealed the national difference in how the lost expedition influenced conventions for Arctic travel writing. In Petersen's hands Rae's report that the Franklin expedition had resorted to cannibalism was used to add dramatic flair, showing that cannibalism was just one aspect of the dangers associated with Arctic exploration.

Like Rae, Petersen had adopted many Inuit ways of life. But he did not need to distance himself from Rae the same way that McClintock did. McClintock, on the other hand, argued that Inuit testimony was always "vague" and could not be trusted, and therefore required the correct interpretation. Commenting on Rae's account, McClintock asserted that his translator, Petersen, had the proper skills to evaluate the truthfulness of Inuit testimony, thus reinforcing damaging Eurocentric stereotypes. McClintock also ensured that the expedition achieved significant scientific results. They carried with them several scientific instruments and established magnetic observatories, and McClintock received training by influential naturalists prior to his departure. This was not unusual for Arctic explorers in general, but this was a search mission. The amount of preparation to maximize scientific research shows how this was still an important focus of Arctic explorations. Significantly, it also functioned as a way to safeguard the explorer against the charges and accusations leveled at Rae.

While Petersen surely would not have disputed McClintock's positive evaluation of his abilities, there was no need for him to construct the same Arctic explorer persona in his narrative. His Danish audience did not share the outrage over Rae's report. When looking at the travel narratives from Arctic explorations in this period, it becomes clear that in the British context science had an important function in the construction of the trustworthy explorer, while these considerations did not extend to the Danish or American contexts. While the Franklin expedition generated new opportunities for Arctic explorations, it also challenged the perception of the explorer in Britain and, as the next chapter will show, transformed the aims and ambitions of empires in the Arctic. The way in which scientific investigations remained central to the outcomes of the search missions—even if this was not explicitly stated in their official orders—shows just how entangled science and exploration was for the creation of authority and trust in the British context. It was a style that heavily favored the stereotypical heroic explorer, one who contributed to a broad range of sciences while undertaking geographical surveys of the Arctic. As the following chapter shows, this was a key stumbling block for the British participation in the First IPY, a venture that aimed at creating a unified Arctic science.

CHAPTER 4

From Science in the Arctic to Arctic Science

The importance of Arctic exploration will again be urged upon the attention of our Government, for the feelings of the people and the press of England cannot now be mistaken. They desire their country to take its ancient place in the van of Arctic discoveries once more.

— Clements Robert Markham, *The Threshold of the Unknown Region*, 1873

The period between McClintock's 1857–1859 expedition in search of the lost Franklin expedition and the First International Polar Year (IPY), also known as the Polar Campaign of 1882–1883, was characterized by a transition in colonial power in the Arctic. The British government was reluctant to send out another expedition to the Arctic after McClintock's. Whereas the British had largely dominated exploration in the Arctic since the end of the Napoleonic Wars, other nations now took center stage. Leading up to and after the purchase of Alaska in 1867, the Americans were stamping their authority there. This transition of power and imperialism draws attention to the resistance and emulation of technologies for traveling and surviving in the Arctic, influencing all aspects of how Arctic expeditions were carried out, from their style to interactions with Indigenous peoples. However, as the secretary of the Royal Geographical Society, Clements Robert Markham (1830–1916), emphasized in the *Geographical Review* and his book on Arctic explorations, *The Threshold of the Unknown Region* (1873), there was a lot of pressure on the British government to yet again assert its dominance in the Arctic: reaching the Geographic North Pole first was a matter of national pride.[1] This recurring and prevailing nationalistic emphasis did not square easily with the intention of establishing international scientific networks, such as the First IPY in 1882–1883. In addition, the portrayal of expeditions as British and American obscured the international composition of these ventures.

Nineteenth-century Arctic science was inherently transnational in nature. Explorers from different nations read and commented upon each other's narratives, and expeditions often included participants from other countries, including Indigenous peoples encountered or employed when there. Would-be explorers could also seek out patrons from multiple countries to fund their ventures. By the 1870s there was a long tradition of international collaborations between individual people and organizations, but the First IPY required a different type of government backing and support. The first of its kind, it was both the culmination and beginning of a shift toward increased collaboration and globalization of nineteenth-century Arctic science. Globalization is a problematic concept, and several approaches to its existence or influence have been suggested in the literature. In many ways, the definition of *globalization* effectively shapes the answer. In "Making Science Global," Marc Rothenberg addressed three ways of conceiving its title. The *global*, he argued, can refer to the breaking down of national boundaries within the scientific community itself—the position that certain scientific questions require a global approach, and as the expansion of Euro-American science to other areas of the world. Jürgen Osterhammel and Niels Petersson's classic text conceptualized globalization as "the development, concentration, and increasing importance of worldwide integration." This is similar to Christopher Bayly's concept of the history of globalizations as the "growing interconnections within the world as such." There is yet another way of understanding global science: Sujit Sivasundaram has suggested a method for global history of science called "cross-contextualization." Here the global indicates the choice of sources and "involves reading across genres and culture." He noted that historians have relied almost solely on European accounts and sources. In addition to considering the use of nontraditional sources, this cross-contextualization also refers to the process of reading the European source within the extra-European context, and vice versa.[2]

Increased international collaborations in the Arctic reflect the globalization of its science from an organizational standpoint, as well as who could textually claim to be authorities on Arctic matters. This shift was not straightforward or uncomplicated, and was associated with transformations in field science, perceptions of the explorer, and the nature and stated aims of Arctic ventures. These tensions were brought to the fore leading up to, during, and following the First IPY. As Markham further wrote in a private letter, "I look upon Weyprecht's scheme [the IPY] as unpractical and of course most injurious to geographical research. He wants men to sit down for a course of years to register observations at one spot, and not to explore. Such as scheme could only have been proposed by the most unpractical of specialists."[3] Explorer versus specialist, geographical research versus "sitting down." This juxtaposition of what Markham (and many like him) perceived as the real purpose of Arctic research with the new vision for science under the IPY is key to understanding the transitions of this period as they relate to both scientific and narrative practices.

This chapter examines three different types of Arctic exploration in a period of transitional imperial powers and scientific methodologies: the British Arctic expedition of 1875–1876 led by George Strong Nares (1831–1915); the Danish expeditions to the interior of Greenland led by Knud Johannes Vogelius Steenstrup (1842–1913) and Jens Arnold Diderich Jensen (1849–1936) in 1876, 1877, and 1878; and the British-Canadian contribution to the IPY at Fort Rae led by Henry P. Dawson.[4] While poles apart in scope and form, these three ventures were each shaped by the transnational and increasingly global nature of Arctic science and the Arctic explorer. Together and in comparison with the second Grinnell expedition and the American contributions to the IPY, they draw our attention to the differences and similarities of how imperial authority was legitimated and practiced, and help to show the varieties of the Arctic experience in this period of transition.

THE FARTHEST NORTH: SUERSAQ'S MANY ARCTIC EXPEDITIONS

In 1875 Suersaq, also known as Hans Hendrik, embarked on an expedition to the Arctic, his fourth as an experienced explorer and translator. As the Danish newspaper *Aarhus Stift-Tidende* reported, he was "the only man who has participated in all the famous expeditions through Smith's Sound."[5] Following the discovery of one of the Northwest Passages, the next big goal in Arctic exploration was reaching the North Pole and finding the hypothesized Open Polar Sea. As I discussed in the previous chapter, Smith Sound was of significant interest in this venture, as several commentators believed that there might be a passageway through the north of the sound which would lead to the Open Polar Sea and the North Pole.[6] The first three expeditions that sought to reach the North Pole through Smith Sound were American: first, the second Grinnell expedition between 1853 and 1855 led by Elisha Kent Kane; second, the expedition led by Isaac

Israel Hayes (1832–1881) in the schooner *United States* between 1860 and 1861; third, the *Polaris* expedition led by Charles Francis Hall (1821–1871) between 1871 and 1872; and fourth, Nares's expedition. Suersaq participated in all of them. He was the first Inuk to publish an Arctic travel narrative, and his account provides important insights into the role of Indigenous peoples employed as part of both European and North American Arctic expeditions. It challenged the accounts of these expeditions constructed by European and Euro-American explorers, especially by those who employed him.

Suersaq was born around 1835 in the small and very poor village of Qeqertarsuatsiaat (Fiskernæs), on the southwest coast of Greenland. He received his education from missionaries, which consisted of the evangelical-Lutheran-inspired Moravian mission (*Mähriske brødre*), also known as the Herrnhuterian mission. The mission was founded around 1720 in Moravia, and three missionaries from the movement arrived in Greenland in 1732 with permission from King Christian VI. In the first half of the nineteenth century, a high-ranking missionary from the group, Konrad Kleinschmidt (1768–1832), had the New and Old Testament translated and published in Greenlandic. In spite of this, the sect did not originally prioritize educating Greenlanders, and their later efforts to do so were not enough to keep the Danish government from transferring their authority to the Danish mission in 1889. The Moravian mission had a detrimental influence on the villages where they functioned. While children were taught to read and write, they were discouraged from learning how to hunt. Suersaq was, however, skilled in both. His abilities did not go unnoticed, and he was recommended to participate in Kane's expedition to search for the lost Franklin expedition on board the *Advance*. When Suersaq's father, Benjamin, passed away in 1852, his family fell on hard times, and Suersaq took over the primary responsibilities of bringing in an income for his family. Suersaq agreed to participate in Kane's expedition because it offered an opportunity to bring in extra funds for his mother, Ernestine, and his brothers. Suersaq was either eighteen or nineteen years old at this point. Kane's expedition was a miserable one, and as Carl Petersen described in his narrative, discussed in the previous chapter, Suersaq abandoned the expedition and settled in Smith Sound, where he met and married his wife, Mequ.[7]

Suersaq's second expedition was led by the American explorer Hayes, who had acted as surgeon during the Kane expedition and knew Suersaq well. This expedition was also plagued by hardship. Their vessel, the *United States*, was damaged by ice, and froze in near Foulke Bay in Smith Sound. The expedition was further troubled by the loss of their dogs to illness. Suersaq's third expedition

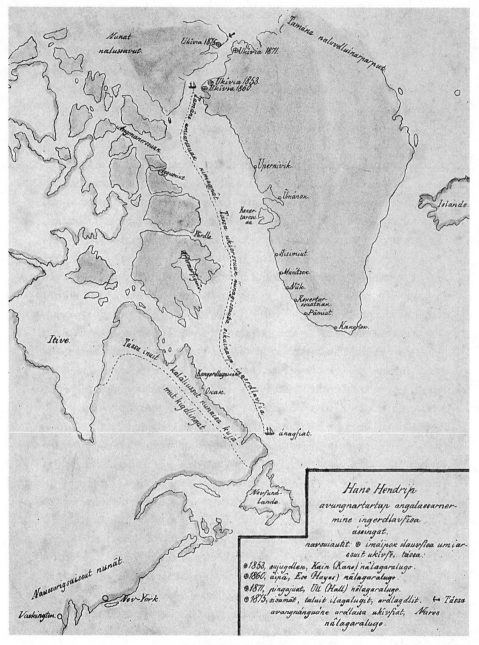

FIGURE 4.1. A sketch map illustrating Suersaq's Arctic travels. The map is assumed to have been drawn in 1878 by Suersaq and Lars Møller. Courtesy of Det Kongelige Bibliotek, Denmark, Shelfmark: KBK 4100–13–1878/2.

FIGURE 4.2. This illustration of Qeqertarsuatsiaat (Fiskernæs) was published both in Elisha Kent Kane's narrative from the second Grinnell expedition and in Carl Petersen's narrative. *Source*: Carl Petersen, *Den Sidste Franklin-Expedition*, 1860, plate 10.

on board the *Polaris* was also dramatic.[8] Charles Francis Hall had command of this expedition, which left New York on June 29, 1871. Hall passed away that same year. The *Polaris* froze in, and although the crew was able to free the ship from its winter quarters by the summer of 1872, the ship was again overtaken by ice in October. Eventually the *Polaris* was so heavily damaged that half the crew abandoned ship with their provisions to seek refuge on an ice sheet. They remained on the ice, including Suersaq and his family, while the other half drifted off with the *Polaris*. The crew left on the *Polaris* managed to get to shore and were later rescued by a whaling ship. The crew on the ice sheet drifted for six months before they were rescued. At this point the ice had drifted close to Newfoundland, a distance of around 1,500 geographical miles. Shortly after their rescue, Suersaq and his family also traveled on board the *Tigress*, which set out in search of the missing members of the expedition who had drifted off with the *Polaris*.

Suersaq's first three expeditions through Smith Sound were American. In this period, there were two main routes used for reaching the North Pole. The German expeditions attempted to reach the North Pole via Spitzbergen, while the American expeditions chose Smith Sound. There was no British route, because after McClintock's expedition the British government had very little interest in further missions to the Arctic. However, in light of the increased American presence there and the very real possibility that the North Pole could be reached by an American expedition, several leading scientific organizations, including the Royal Society of London, the British Association for the Advancement of Science, and the Royal Geographical Society, and public figures began to put pressure on the government. The Dundee Chamber of Commerce, which represented the interests of Scotland on this matter, also drew up a recommendation to the government. On December 17, 1872, representatives from the Arctic committees met with Chancellor of the Exchequer Robert Lowe (1811–1892), and First Lord of the Admiralty George Goschen (1831–1907). The deputation included Arctic veterans such as George Back; the president of the Royal Geographical Society, Henry Rawlinson (1810–1895); and the prominent scientists Joseph Dalton Hooker, John Lubbock (1834–1913), and William Spottiswoode (1825–1883). Rawlinson read a letter that was later reprinted in British newspapers and journals. As with the search for the Northwest Passage and the Franklin expedition, the quest for the North Pole was repeatedly framed as a matter of national pride for Britain. These concerns were present in the 1860s and 1870s, as they had been in earlier periods: who held the territorial dominance in the Arctic was a key motivating factor.

As Rawlinson wrote in his letter, "The belief is expressed that all classes of the people will unite with men of science in the desire that the tradition of Arctic discovery should be preserved and handed down to posterity, and that Englishmen should not abandon that career of noble adventure which has done so much to form the national character, and to give our country the rank she still maintains."[9] In addition to playing up the nationalistic concerns, Rawlinson's letter to the government listed the support of several major scientific societies: the Royal Geographical Society, the Royal Society, the Geological Society, the Linnaean Society, the Scottish Meteorological Society, the Metrological Department, the Anthropological Institute, and the British Association for the Advancement of Science. The associated groups show the broad scope of scientific disciplines that were interested in the potential results of Arctic explorations, positioning such missions as being of interest and value to all scientific fields. Rawlinson further emphasized this point by outlining a highly ambitious list of possible scientific results of an expedition to the North Pole. In practical terms, the deputation suggested that the government send out two whaling ships, each carrying sixty men. The proposed expedition should start in May and plan for two winters in the Arctic. Like the Americans, this proposed expedition would go up the west coast of Greenland, as they could make use of the Danish settlements for aid if needed. However, not everyone agreed with Rawlinson's evaluation.

When the Arctic Committee deputation published their recommendations, a highly critical editorial appeared in the *Times*. Likely written by the paper's editor, John Thadeus Delane (1817–1879), it described the promised scientific results from such as venture "as unexpected, striking, and sensational as any Christmas literature may be thought to require." However, he argued, as scientific societies want experiments and observations, and Navy men want employment, they would naturally be inclined to propose such an expedition. Like the Northwest Passage, which was "a pure 'phantom of the scientific brain,'" reaching the North Pole would require an expedition "by an unknown route, to an unknown region, with purposes which are not only hopeless because they are unknown also." Another point of view was expressed in the same newspaper by an anonymous author, who signed his name as "An Arctic Officer." The author described himself as a veteran of Arctic explorations as part of "what I may be pardoned for calling this glorious stage of the Arctic drama," and countered the lead published in the *Times* to echo Rawlinson's points. Yes, the Arctic Officer argued, naval men desire employment on Arctic expeditions for the associated glory and excitement, but this was not the primary motivator for advocating an expedition to the North Pole.[10] There was immense value to a new Arctic expedition for naval training, and any concerned reader could rest assured that Arctic expeditions were much safer now.

John Rae disagreed with this estimation and aired his reasons in a letter, also in the *Times*. While the Arctic Officer claimed that the lead in the *Times* was the only one in the daily and weekly newspapers who "has thrown a damper on our hopes" for a British expedition to the North Pole, Rae offered a third point of view. While the Arctic Officer claimed that the "people of Hull, Dundee, Aberdeen, &c." confirmed it was possible to "steam round Smith's Sound" in one summer, Rae argued this was just as unbelievable as if a "naval seaman" had told him that "he could with certainty bring his ship safely through the most intricate navigation on our coasts (of which he had no previous knowledge) in a dark night, without the aid of soundings, chart, or compass." Although Rae had faced heavy criticism after his report on the fate of the lost Franklin expedition, he had arguably charted more coastline by sled and snowshoe than any other Arctic explorer. Even Jane Franklin stated, though reluctantly and privately, that

Rae was one of the best overland Arctic explorers. Rae's rejection of the Arctic Officer's claim, that it would be no problem to reach the North Pole by sled as "greater distances had already been accomplished," is likely to have carried some weight.[11] He argued that the Arctic Officer had no way of determining the distance they needed to travel overland, as they could not predict where the ship would winter.

It is important to emphasize that Rae did not argue against an expedition to the North Pole itself. Rather he strongly advised the government not to follow the route through Smith Sound, which had been suggested by the Arctic Committee deputation. Based on the experiences of American expeditions led by Kane and Hayes, Rae believed another through Smith Sound was a very bad idea. Instead he advocated for one of "more humble pretensions," via Spitzbergen.[12] Rae was not the only dissenter. John C. Wells, a captain in the British Royal Navy and author of *The Gateway to Polynia*, wrote several letters to the *Times* also advocating for a British expedition to the North Pole via Spitzbergen.[13] Wells believed that the North Pole could be reached in one season via this route by use of a steam whaling ship. This would allow the explorers to reach a higher latitude by ship than they could in Smith Sound, and they would then be able to travel by sled to the North Pole.

The cost of previous Arctic expeditions, both economically and in human life, was a significant reason for many others to argue against new expeditions. In fact, Chancellor of the Exchequer Robert Lowe and First Lord of the Admiralty George Goschen were not fully supportive of Rawlinson's proposal for an expedition either. The British government sent out an expedition with the HMS *Challenger* between 1872 and 1875, which was very expensive. As Lowe and Goschen replied in a letter to Rawlinson, "Under these circumstances, we regret that we cannot recommend the sending an Exploring party to the Arctic Ocean as a Government Enterprise this year."[14] Economic factors were not the only consideration. As the historians Marvin Swartz and Frank Herrmann have noted, the question was not only what economic value colonial possessions could bring, but also what value there was in not letting potential colonial possessions fall into the hands of their rivals. Because of this, they argued that the prime minister "nervously watched the entire globe."[15]

By 1874 Benjamin Disraeli (1804–1881) had returned as prime minister, and he was very keen to reassert British dominance in the Arctic. In 1874 Disraeli wrote to Rawlinson that, after "having carefully weighed the reasons set forth in support of such an expedition, the scientific advantages to be derived from it, its chances of success, as well as the importance of encouraging that spirit of maritime enterprise which has ever distinguished the English people," his government had determined to organize an expedition to the North Pole through Smith Sound. The result was the Nares's expedition of 1875–1876. Although the first three American expeditions by all accounts were horrible, Suersaq agreed to participate in the British venture. However, this time he did not bring his family with him and described how "as I was now going to depart, I pitied my wife and my little children who were so attached to me."[16] He had brought along his wife, Mequ, and his children on both his second and third expeditions. One of their children, in fact, had been born during the Hall's expedition and was given the name Charles Polaris after the late Hall and the wrecked ship. It is uncertain, though not unimaginable, why Suersaq did not bring his family with him on this fourth voyage through Smith Sound.

Despite Rae's warning that any mission to the North Pole should be scaled down, the organization of Nares's voyage followed the typical British style. While Rae had shown the importance of adopting Indigenous methods for traveling in the Arctic, such as the use of snowshoes, the Admiralty evidently disregarded these recommendations. Even in the face of repeated disasters, they stuck to sending out one or two large ships, fitted with lots of provisions, a large crew, and expensive scientific equipment, which had been shown to be ill-suited for Arctic exploration. This was a serious mistake, and the expedition struggled with heavy sleds, small tents, and the effects of scurvy.[17] Following his return home, Suersaq was encouraged to write a memoir of his experiences by the inspector of North Greenland, Sophus Theodor Krarup-Smith (1834–1882). Suersaq's account was published in serialized form in the Greenlandic newspaper *Atuagagdliutit* between September 1878 and January 1879. The full narrative was published as *Memoirs of Hans Hendrik, the Arctic Traveller, Serving under Kane, Hayes, Hall and Nares, 1853–1876* (1878). It was translated from Greenlandic to English by the director of the Royal Greenland Board of Trade, Hinrich (Henry) Rink (1819–1893).

Nares made several brief references to Suersaq in his own narrative and in his *Official Report of the Recent Arctic Expedition*, where he provided the following biography: "All speak in the highest terms of Hans, the Esquimaux, who was untiring in his exertions with the dog-sledge, and in procuring game—it was owing to his patient skill in shooting seal that Dr. Coppinger was able to regulate the diet somewhat to his satisfaction." From this, it would appear that Suersaq functioned primarily as a hunter and dogsled driver. During the winter period,

Suersaq described how he was left largely without anything to do, as he "was not engaged for sailor's work, but only as hunter, sledge-driver and dog feeder. This is what I had promised on leaving my home."[18] However, the portrayal of events in his narrative complicates this perception and shows the tensions surrounding the acceptance of Suersaq's authority as a firsthand observer of the Arctic.

The difference between the portrayal of Suersaq's role in his and Nares's narratives is particularly striking when we consider the extent to which both described how the expedition relied on Suersaq's skills to carry out their overall goals. For example, while the ships were lodged in ice, Lieutenant Lewis Beaumont was in charge of a team to survey the north coast of Greenland around Polaris Bay. The crew began to suffer from the effects of scurvy, and they organized a return party that, according to Nares's *Official Report*, was "helped by Hans." Nares gave more credence to Suersaq in his personal narrative, where he noted that it was "mainly due to Hans' clever management of the dogs, and his skill as a driver, that we were enabled to advance so rapidly with such a heavy load."[19] Suersaq went into further detail and described how he transported the four sick men, going back and forth with two at a time. They relied on him, Suersaq noted, to find the safe routes through the snow and ice. Suersaq described an exchange between Nares and him as follows:

> After having stayed some days, he said to me "Tomorrow we will repair to the ship; go with us as our guide." Next morning we went off, leaving the sick, who had begun to walk about. When we were going, our Captain said: "Now, show us the road; go ahead of us, and we will follow." Thereupon we started, and crossed the open water in a boat. When we came to the heavy ice, I searched for the best road, accompanied by the Captain. He used to question me: "Which way are we to go?" I answered: "Look here; this will be better." It was lucky the Commander treated me as a comrade."[20]

The expedition further relied on Suersaq's help in employing other Indigenous peoples to support the expedition as they progressed north. For example, Suersaq noted that it was his choice to visit a settlement by Ivnanganek (Cape York) "and try to find the man [Augina] I wished to take along with me." They were unable to locate Augina, yet Suersaq's account shows that his participation in three previous expeditions had provided him with an intimate knowledge of all aspects of the region and that Nares relied on him to make decisions about the hiring of additional crew. Nares did not explicitly mention how Suersaq made decisions on whom to employ and made no mention at all of Suersaq's support within the context of the geographical surveying accomplished by the expedition. This is, however, evident from Suersaq's accounts, as he described how "when bright daylight had set in, the Captain and I used to travel about by sledge, to measure the height of the mountains." During the three winter months, he "also did duty as the Captain's sledge-driver in surveying the country and climbing the hills." This reveals that Nares relied on Suersaq to fulfill his duties in geographical surveying and collecting the natural history specimens the expedition was expected to bring back to England. In addition, Suersaq surveyed the land independently, noting that "when he [Nares] remained at home, I went alone," yet this contribution was rendered invisible in Nares's accounts.[21]

The issue of Rink's translation raises several questions. First and foremost, how true was the translation to the original? Scholars such as Julie Cruikshank have emphasized some of the problems associated with translating or retelling oral narratives from their original languages to English. These differences relate to more than the explicitly apparent grammatical issues and narrative conventions, such as sentence structure and gender distinctions; it is important to remember that the plurality of spoken languages was a key feature of the transnational nature of Arctic expeditions. The main language of an expedition may have been English or Danish, but these were often second or third languages to the crew members and those encountered in the Arctic. As the Inspector Krarup-Smith noted in a private letter, "mingled languages" were common in Greenland and could be the cause of misunderstandings and confusion.[22]

Of equal significance to language are the subtle differences in tacit knowledge and assumptions. In *The Wretched of the Earth* (1961) Frantz Fanon divided the writings of colonized peoples into three stages. In the first, "imitative" stage, the literature copies the form of the colonizers. The second stage rejects these paradigms, while expressing a nostalgia for the perceived authentic Indigenous. The third "fighting" stage rejects the literature of both the first and second stage to create a new democratic, postcolonial literature and culture. Drawing upon this, Michael Wilson has examined the linguistic and stylistic form of Native American literature and argues that genre and stylistic choices were not neutral, but an expression of a hierarchical relationship between the Indigenous and imperial practices. Wilson's approach sees literature as a dialogic exchange, with different levels of resistance, emulation, and novelty. Suersaq's narrative is reflective of such an exchange.[23] While *Memoirs of Hans Hendrik* was constructed as a travel account, it was stylistically very different from those published by British, Danish, and Euro-American explorers.

Nares's *Narrative of a Voyage to the Polar Sea* followed the standard day-to-day travel narrative format. It was published in two volumes and included a large appendix with extensive scientific results. It also made use of the well-known tropes of the heroic Arctic explorers being full of drama that emphasized the dangers of the region. By contrast, Suersaq's *Memoirs* was almost completely void of this type of rhetoric. It was a retrospective account of his expeditions, and the chronology of events was not clearly demarcated. Rather, each of the four expeditions was recounted as a story. The chronology of Suersaq's narrative is one area where Wilson's and Fanon's notion of resistance and emulation becomes clear. Suersaq listed dates primarily when he recounted the number and type of game he had caught, important information for hunters. Yet instances that would appear of much greater significance, such as when he temporarily abandoned the ship, were dated only with reference to "the dark season."[24] Rink himself offered as explanation in his introduction to the *Memoirs* that Suersaq had kept few written notes from his expeditions. These notes were said to have only briefly described the country they surveyed and its inhabitants.[25] Accordingly, the majority of the *Memoirs* was likely compiled retrospectively from memory.

Rink made several editorial changes to Suersaq's text in his translation. Although he noted that "the manuscript is written in tolerably plain and intelligible Greenlandish," there were "some words here and there [which] remained inexplicable or doubtful, and some sentences unclear." In the body of the text, Rink indicated those instances of translational uncertainty. He also claimed to have maintained Suersaq's spelling of personal names. For their contemporary readers, this stylistic choice added an air of authenticity to the memoir, but it also reveals significant aspects about naming.[26] Naming people and places was (and is) part of the ideological challenges from and to colonialism, as is evidenced in recent years in the changes to official place names from Danish to Greenlandic. It was also used by reviewers to illustrate Suersaq's low educational level, something that was not likely to have been Rink's intention.

Rink was strongly committed to advancing the living standards of Greenlanders, and he encouraged the systematic surveying and cataloging of Greenland's natural history and the publication of the results beyond Danish borders. He received his early education at Sorø Akademi and his doctorate at the University of Kiel. He acted as the geologist for the Danish expedition to circumnavigate the world on board the *Galathea* between 1845 and1847. Afterward he took up several high-ranking administrative positions in Greenland. In 1853 he married Nathalie Sophie Nielsine Carlonie Møller (1836–1909), known as Signe Rink, who was born and raised in Greenland, the daughter of the colonial administrator in Paamiut, Jørgen Nielsen Møller (1801–1862). She published several books and articles about Greenland, in particular on the subject of its ethnology, and she translated several books. Together Rink and Signe were part of founding the first newspaper in Greenland, *Atuagagdliutit* in 1861, which was published in Greenlandic.[27] They both believed they could improve the lives of Greenlanders—although it is important to note that this belief was embedded within the civilizing and colonial rhetoric of the time.

Rink's obituary reflects upon the way he perceived himself in opposition to the KGH.[28] The company, Rink argued at several points, did not work with the interests of Greenlanders in mind. As he was director of the KGH between 1871 and 1882, this criticism was particularly poignant and was part of a very public controversy between Rink and other influential figures in the company, which eventually led to Rink's retirement. In 1877 Rink gave a talk to the Royal Danish Geographical Society (Danske Geografiske Selskab), where he described how the general well-being of Greenlanders had deteriorated significantly during the previous thirty years. One key problem was, according to Rink, that foreigners in Greenland had not appreciated how Greenlanders perceived the cultural differences between them. Rink further addressed this issue in his introduction to Suersaq's memoir, stressing that the horrible ways Europeans had treated Greenlanders naturally influenced how they interacted with them. Rink wrote that when foreigners came to Greenland with the attitude that the Indigenous population was inferior and could only communicate through interpreters, it was no surprise that they "at times must feel himself misunderstood and wronged."[29] This explained many of the instances of conflict, misunderstandings, and distrust between European explorers and Inuit hired to participate on the expeditions.

One particularly distressing episode of conflict was described by Suersaq, but not recollected in Nares's narrative. During winter Suersaq repeatedly heard other crew members talking about him. One evening he overheard them plan a physical assault on him, "'When Hans is to be punished, who shall flog him?'" This was not the first time Suersaq had been threatened, and he was concerned enough to leave the ship as "if I should freeze to death it would be preferable to hearing this vile talk about me." It was nighttime when Suersaq left, and he walked about five miles before turning back to sleep in the snow close to the ship, in the hopes that Nares would search for him. Nares did send out a search party and told Suersaq to let him know if the other crew members repeated their threats. Yet the harassment of Suersaq did not cease, and he wrote that "I afterwards heard them speaking several times in the same way, but, nevertheless, did not mention it, because I supposed that, if I reported it, none of them would like me more." A review of the *Memoirs* in the *Athenaeum* commented upon the implications of this event, noting that "it is indeed not a little humiliating to find that he was always in terror of being flogged, both on board the American and English expeditions."[30] The account reveals not only the isolation Suersaq felt during the expedition, but the derogatory way he was treated by some of the crew. It shows what the contributors to *Implicit Understandings* (1994) describe as "implicit ethnography." The comparison with Nares's *Narrative* and Suersaq's *Memoirs* illustrates how there was a tacit process of conceptualizing oneself and the other on both sides of the encounter between foreign explorers and Indigenous peoples. Suersaq himself wrote that he did not know if he had "thoroughly understood their meaning" and had no "particular purpose" for writing about their intentions to flog him.[31] Yet his account and Rink's decision to keep it in the published narrative were very poignant critiques of European colonial power in Greenland and provided a window into how European explorers treated their Greenlandic employees. As Rink further argued, the long history of the way Europeans had behaved in Greenland affected Suersaq's perceptions of the explorers he worked with: "However, thoroughly to understand the strange suspicions exhibited in some parts of his statement we must consider the traditions still living amongst the Greenlanders about atrocities formerly committed in their country by foreigners, as well as their indistinct ideas of the wars and military discipline of the white men."[32] Suersaq's narrative was accordingly a highly politicized piece as it challenged the accounts of previous explorers from other nations and problematized past and present imperial policies in Greenland.

Rink emphasized that Suersaq's narrative was trustworthy even when his account of events differed from his captains'. Suersaq had an excellent memory, Rink argued, and as he had not read the other narratives from the expeditions, he was not influenced by their accounts. A review in the *Athenaeum* agreed, and further noted that it was "probable that [Suersaq's] sketch of Hall's expedition is on the whole more trustworthy than any other we possessed until recently." The *Athenaeum* further noted that his narrative was "not only quaint, but really valuable ... both from an historical and ethnological point of view." Rink's decision to translate and publish the memoir was clearly influenced by his political ambitions and commitment to his version of humanitarianism, and should be seen within this context. Similarly, Krarup-Smith, the inspector of North Greenland who encouraged Suersaq to write his memoir, had concerns about the behavior of American and British explorers and traders. After the disastrous end to the *Polaris* expedition, Krarup-Smith wrote a letter (which was never delivered) to Daniel Lawrence Braine (1829–1898), an admiral in the US Navy and commander of the *Juniata*, which had been part of efforts to search for the missing *Polaris* expedition. "I have been very sorry to hear, that the Esquimaux Hans Hendrik very nearly had deserted the 'Tigress,'" Krarup-Smith wrote, because he had "been pursued by suspicions" as well as "even frequent threatening which made his life on-board almost insupportable."[33]

Another important aspect of the publication of the *Memoirs* is Rink's scientific research program, including the dissemination of knowledge to an international audience. While large extracts were published in Danish in the *Geografisk Tidsskrift* in 1877, the full memoir was published only in English. Rink translated Suersaq's *Memoirs* into English to give it a broader audience, but it also meant that he was translating into a language that was not his own. Therefore an English-born professor of English at Copenhagen University, George Stephens (1813–1895), edited it. Stephens's involvement in Suersaq's narrative poses something of a conundrum. In his introduction to the narrative, Rink outlined Suersaq's many accomplishments and his extraordinary abilities. For example, he wrote that "our author affords a striking example of the independence of his nation, of the climate within their vast territories, as well as of aid from foreign nations."[34] By contrast, Stephens somewhat undermined Suersaq's authority in one swift brushstroke by stating in his editorial translator's note that he "thought it best to let Hans Hendrik write in the naive way to be expected from such a child of nature." This racist rhetoric was countered by a feeling of truthfulness in Suersaq's observations of the behavior of the European and Euro-American men he had traveled with, and an acknowledgment that his geographical and ethnographic observations were valuable. The review that appeared in the *Examiner* is a good example of such racist rhetoric, and of the tension it created about Suersaq's authority as a firsthand observer of the Arctic. The review noted that "a literary composition by a pure-blooded and unsophisticated Eskimo must always be interesting," but that Suersaq had a "reputation of being the most truthful individual."[35]

Judging by the reviews, many readers would assume that Suersaq was uneducated. For example, the *Athenaeum* noted that his education would "not allow of many rhetorical flourishes" and described his *Memoirs* as a "quaint, simple narrative, with all its blunders in orthography, geography, and nomenclature, bears the obvious marks of stern fidelity to the truth."[36] However, when Suersaq was a child the main criticism directed at the Moravian mission was for discouraging Greenlandic children to learn how to hunt, more so than for their lack of scholastic efforts. Suersaq spoke Greenlandic, Danish, and English, and he could read and write. He was a trusted guide both for navigating the icy landscape and in negotiating the assistance of Indigenous peoples along the way, so much so that the captains of four expeditions deliberately chose him for their missions. In addition, Suersaq was raised as a Christian, and his parents assisted the clergy at the Moravian mission. Readers of his narrative should have been compelled to accept his word as a truthful representation of the expeditions and the Arctic, yet the reviews still positioned him as inferior. Variations on describing Indigenous peoples in Greenland and North America as "children of nature" were present in much of the literature about the Arctic. As Chapter 2 in particular showed, the rhetoric of the civilizing mission in Greenland combined the conversion of the Indigenous population to Christianity with an acute sense of imperial superiority—they were brothers in Christ, but not equal.

Suersaq was highly skilled and had written a narrative that in some instances corrected the information in other travel narratives from the expeditions he participated in—and he exposed the dark side of how European explorers treated Inuit. While Suersaq's *Memoirs* was written in the familiar format of a travel narrative, it broke with stylistic conventions on several fronts, as shown throughout this section. In this period of increased international collaborations in the Arctic, Suersaq's narrative fit uncomfortably into the category of accepted sources about the Arctic for his contemporary readers. As a go-between, Suersaq had insights into Inuit, British, Danish, and Euro-American cultures. However, his expertise was not easily accepted, and the difference in stylistic and narrative structure of his *Memoirs* was used against his authority. From his *Memoirs* it would appear that Suersaq did not know the names of the expedition captains. For example, "Tartikene" referred to Doctor Kane, and "Tart Eise" to Doctor Hayes. The reviews picked up on this as an illustration of Suersaq's poor language skills. However, whereas the correct or exact naming of people and places was a way to show accuracy in travel narratives as a scientific document (recall how the Croker Mountains haunted Ross's career), Suersaq evidently assumed the reader would know who he was referring to. Rink's translation from Greenlandic to English is a key issue here, as is the translation from oral story to written text.

It is possible that there were misunderstandings about the use of nicknames, or phonetic spelling of names, that Suersaq had likely transliterated their names differently. Without Suersaq's original manuscript, it will remain guesswork. What is certain though, is that it was used as evidence for Suersaq's lower social and educational status. The tension between accepting Suersaq as an authority and still describing him as a child of nature shows the precarious role Indigenous assistants to Arctic expeditions held. As a cultural intermediary, he evidently fit uncomfortably into the perception of Inuit and challenged the notions of who constituted an authoritative Arctic writer.

SCIENCE, FINANCE, AND DANISH IMPERIAL AMBITIONS IN GREENLAND

The acceptance or rejection of Suersaq as an authority on the Arctic connected to the complex perceptions of who was a trustworthy observer, and his narrative hints at the global nature of Arctic science and the Arctic as a contact zone. Rink was a central figure in the publication of Suersaq's narrative, and his ambitions to control the direction of Arctic research influenced Arctic science beyond his Danish national context. Rink's research program was shaped by his international network of research affiliates, as well as the ethos of Denmark after the First and Second Schleswig War. The saying "For every loss a replacement can be found, what has been lost outwardly must be regained inwards," coined by the author H. P. Holst (1811–1893), came to symbolize the mood in Denmark after the Second Schleswig War.[37] What had been lost outwardly could be regained through intensified industrial and scientific effort. This included exploration of Greenland, and the first volume of *Meddelelser om Grønland* (est. 1879) included several discussions on the possible monetary value of increased extraction of minerals in Greenland. As was noted in the third volume, "It would seem natural that Greenland, of which the biggest and most thoroughly examined areas belong to the Danish Monarchy, is also described through Danish efforts in its Botanical aspects, as it presumably has been described by Danish scientists in other areas."[38]

The Commission for the Direction of Geological and Geographical Investigations in Greenland (Kommissionen for Ledelsen af de Geologiske og Geografiske Undersøgelser i Grønland) organized a series of expeditions to survey the interior of Greenland starting in 1876. The first was led by the geologist Knud Johannes Vogelius Steenstrup and the geologist and naval officer Jens Arnold

Diderich Jensen between 1876 and 1878. Steenstrup, who was nephew to one of the most influential Danish scientific figures of his time, the zoologist Japatus Steenstrup (1813–1897), had already carried out geognostic examinations in Greenland in 1871, 1872, and 1874 and had worked as a museum assistant at the Mineralogical Museum in Copenhagen since 1864. Results from the expeditions were typically published in *Meddelelser om Grønland* or as travel reports.[39] While the prospect of an economic payoff was a central factor in the increased interest in exploring Greenland, there was also a not-insignificant level of national pride associated with these expeditions, hence the emphasis on it being a Danish endeavor. In many ways, *Meddelelser om Grønland* was a representation of a trinity of science, finance, and national pride.[40]

Just as *Flora Danica* was available in libraries and folk high schools (*Folkehøjskoler*), as discussed in Chapter 1, *Meddelelser om Grønland* was linked to an ethos of knowledge dissemination. As part of the Commission for the Direction of Geological and Geographical Investigations in Greenland, Rink played a key role in establishing *Meddelelser om Grønland* in 1879. While Rink chose to direct his publications to the Danish- and English-speaking audiences, *Meddelelser om Grønland* was originally addressed to Danish- and French-speaking readers. Later volumes were fully translated into English, but throughout the nineteenth century English was a marginal language in Denmark, with French, German, and Latin as the main languages of elite education.[41] The goal of *Meddelelser om Grønland* was both to catalog the Danish Empire and make the knowledge available to a broad audience—including researchers from other countries. Danish researchers were increasingly working in collaboration with people from other countries, including Sweden, Norway, the Netherlands, France, England, and Germany. This is reflected in *Meddelelser om Grønland*, which regularly included foreign language abstracts. As was noted in volume two, "As the Danish language is not broadly understood, we have tried to make up for this by accompanying every issue of *Meddelelserne* with a French abstract, as it in this way does not lose the character of being a Danish endeavour."[42]

Paradoxically, the national pride expressed over the Danish efforts to colonize and survey Greenland was coupled with a peculiar sense of being the underdog. The Danish colonies in India and Africa were lost or sold in the middle of the nineteenth century, but Denmark maintained colonial power in other parts of the world, including what is now known as the US Virgin Islands. As was noted in the introduction to the first volume of *Meddelelser om Grønland*, anything Denmark achieved in Greenland was done with means that were limited compared with those of other nations, as "there has been made efforts to adjust them after our own situation, and that no larger project was begun, before it was possible to complete them."[43] This perception of having to justify any costs associated with Arctic ventures was reflected in the style of the expeditions organized. Small and cheap, the goal was to survey as much as possible.

While the coastline of Greenland was slowly being charted, the interior of the country was completely unknown. The first expedition to succeed in crossing Greenland was led by the young Norwegian explorer and scientist Fridtjof Nansen (1861–1930). Nansen's expedition, consisting of six men, traversed the ice sheet on skis from the eastern to the western coast in 1888. The choice of starting in the east, rather than from the Danish settlements in the west, broke with the plans of previous missions. The reasoning behind this decision was that, if problems were encountered, turning back would not be an option for them as had been the case before. Starting on the east coast meant that there were no settlements to return to for relief, thereby forcing them to complete their goal. In the 1870s, however, attempts at penetrating the interior started at the Danish settlements on the western coast. One reason for Nansen's change was that the attempts to transverse Greenland in 1877 and 1878 were only one aspect of the expeditions' goals. Covered by a seemingly unending ice sheet, also known as the inland ice (*indlandsis*), the yet-impenetrable interior was a source of mystery. What it could reveal about past ice ages and what was hidden under the ice were key topics of interest. Rink and his small group of Danish colleagues, influenced by the methods of Abraham Gottlob Werner and Karl Ludwig Giesecke (1761–1833), believed that the forces at work in the frozen north were the same that had shaped Europe during the last ice age.[44]

In 1876 Steenstrup charted the area around Qaqortoq (Julianehaab) together with the geologist Andreas Nicolaus Kornerup (1857–1881) and naval officer Gustav Frederik Holm (1849–1940). Steenstrup had previously carried out geognostic research in Greenland, and so was a strong choice to lead the expedition. Its primary aims were to carry out geognostic and geographical research of the area, but also to undertake preliminary examinations of the border of the ice sheet. Because Steenstrup was very familiar with the area, the expedition was able to survey and produce a geognostic map of a very large 4,000-km^2 area. There is an interesting difference in the language used between the description of the expedition as it appeared in the first volume of *Meddelelser om Grønland* and in the 1912 overview in the journal *Oversigt over Meddelelser om Grønland*. The first volume used the terms *geognostic* (*geognostisk*) and *geological* (*geologisk*)

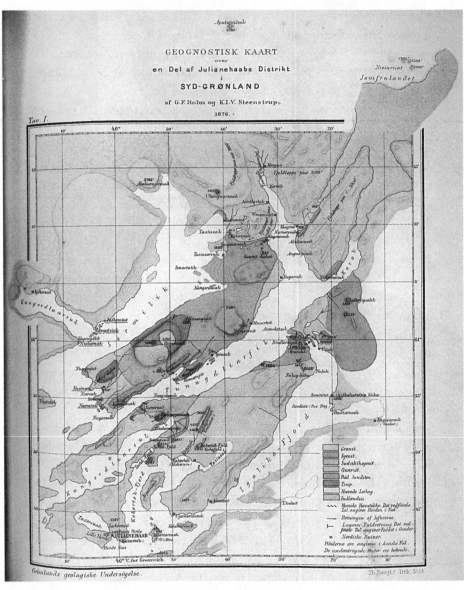

FIGURE 4.3. A geognostic map of Greenland, in full color. *Source: Meddelelser om Grønland*, 1879, vol. 1, plate 1. Courtesy of the Scott Polar Research Institute.

interchangeably. In 1912 *geognostic* was used to refer only to Steenstrup's map from 1876. The difference between geognosy and geology is subtle but significant. The term *geognosy* was coined by Werner to refer to a science distinct from

FIGURE 4.4. An illustration showing the parallel cleaves in the inland ice, published in full color. *Source: Meddelelser om Grønland*, 1879, vol. 1, plate 2. Courtesy of the Scott Polar Research Institute.

natural history, mineralogy, or geology. He believed that the term *geology* was used by speculative writers, and introduced geognosy to differentiate a science of the Earth that is firmly based on empirical evidence. As discussed in Chapter 3, Werner's theories about the Earth and his methodology were highly influential. Steenstrup worked within a variation of the Wernerian notion of universal formation of the different physical characteristics of stratigraphy for his research in the Arctic, just as Richardson did. The "Wernerian radiation," as coined by Rachel Laudan, extended widely.[45] Werner's influence appeared throughout the entire institutional infrastructure of geology on the European continent, but by the early twentieth century *geology* had become the standard term.

Another person particularly significant in shaping the geological examinations of Greenland in the nineteenth century was the German mineralogist and explorer Karl Ludwig Giesecke. An 1801 visit to Werner in Freiberg left a deep impression on Giesecke. The historian Alexander Whittaker has argued that this trip "was particularly important in demonstrating how [Giesecke's] Greenland scientific work and results managed to be fully up to date within the prevailing geological paradigm, not only in terms of Werner's mineral system, but also within the developing Wernerian ideas on geognosy and geological sequence."[46] Giesecke had close ties with the Danish geological community. He lived in Copenhagen as a mineral dealer and traveled to Greenland by royal request to undertake a survey of the country's mineral wealth. Giesecke's study of Greenland's mineralogy was hugely influential, especially so among the founding figures of *Meddelelser om Grønland*. For example, in 1878 the Danish geologist Johannes Frederick Johnstrup (1818–1894), professor of mineralogy at the University of Copenhagen and editor of *Meddelelser om Grønland*, published Giesecke's diary with a supplement by Rink. Johnstrup and Steenstrup also published an updated edition of Giesecke's diary in 1910.[47]

What was under the ice sheet covering Greenland? Giesecke's "Remarks on the Structure of Greenland in Support of the Opinion of Its Being an Assemblage of Islands, and Not a Continent" influenced scientific understandings of Greenland's interior and as such the decisions to send expeditions in search of the North Pole through Smith Sound. Giesecke's paper, which he had sent in a letter to Scoresby, outlined his viewpoint that Greenland was not a continent but consisted of several islands bound together by ice. The implication of the paper was that Greenland's connected islands extended into the hypothetical Open Polar Sea, later used by Kane and Hayes to support their Smith Sound route in search of the North Pole.[48] These two theories were persistent. As the Arctic geographer Robert Brown (1842–1895) proclaimed in 1875, "Greenland has no Interior! at least if we look upon its interior in the light of something else than ice and snow." Rink, however, was not as willing to speculate, simply noting, "Wherever one attempts to proceed up the fjords of Greenland, the interior appears covered with ice; but there is no reason whatever to assume that this applies to the central part of the country, in which one, on the contrary, just as well may assume that there are high mountain chains, which protrude partly from the ice." And yet, in the book *Danish Greenland, Its People and Its Products* (1877), which incidentally was edited by Brown, Rink noted that the interior of Greenland was made up of "numerous islands throughout the whole of its extent."[49]

In 1877 Steenstrup and Jensen examined the northern part of the Frederikshaabs District. Because of particularly rough weather, they were unable to travel any meaningful distance into the Greenland ice sheet, but in spite of this the expedition generated significant scientific results. In 1878 Jensen traveled back to Greenland as the leader of an expedition to the southern coastline. He was assisted by Kornerup and the architect and painter Ernst Thorvald Groth (1847–1891). His official instructions gave Jensen free rein with regard to the route and delegation of tasks. The hope was that Jensen's expedition would be able to survey the coastline from the mountain Tiningnertok to the Ameralik Span. Scientifically the focus was on "all aspects of the physical geography" and "archaeological observations," as well as specific features of the ice.[50] They charted a large stretch of coastline and prepared a geological and topographical map. They also estimated the height of nearby mountains by means of trigonometric calculations and a barometer. The collection of minerals to determine the possibilities for mineralogical extraction was also a central part of the expedition. Notably the 1878 expedition brought home over 1,000 plants in 120 varieties, including 27 varieties on the Jensens nunatakker alone.

The Jensens nunatakker was discovered and named during the expedition in 1878, when they succeeded in penetrating seventy kilometers into the ice sheet. This was significantly further than had done before.[51] The nunataks (or *nunataqs*, ice-free peaks in the ice sheet) were of particular interest for several reasons. They indicated what was under the ice sheet and made it possible to study the motion and behavior of very large bodies of glacial ice. The ice moved around and against nunataks, and the pressure of the ice against the rock shaped the glacier and created terminal moraines. The processes that had shaped the landscape during the ice age could be observed from the vantage point of the nunataks. The dynamics of glacial movement was here of a different magnitude than where it had been studied in Europe.[52] Jensen also published his findings from the 1878 expedition in the journal *Geografisk Tidsskrift*. Where *Meddelelser om Grønland* had focused on the scientific results of the expedition, this article also described their experiences of surveying, with particular emphasis on the dangers associated with penetrating the inland ice.[53] The article included several images drawn by Kornerup, which in addition to illustrating the phenomenon of glacial fractures, also made it clear why crossing the interior of Greenland was so dangerous.

The expeditions in the second half of the 1870s were considered "a type of trial year for the examinations in the area of Greenland where the Danish colonies are placed," and both *Meddelelser om Grønland* and *Geografisk Tidsskrift* urged the government to fund more expeditions, arguing that Denmark had a special obligation to carry them out. It was research that would not be possible without government support and funding, even if "others in foreign countries have felt the absence of it, and, at least for the first part, complained that Denmark had not fulfilled their obligations in this area long ago."[54] The belief that Danish-led research in Greenland should feature within the international

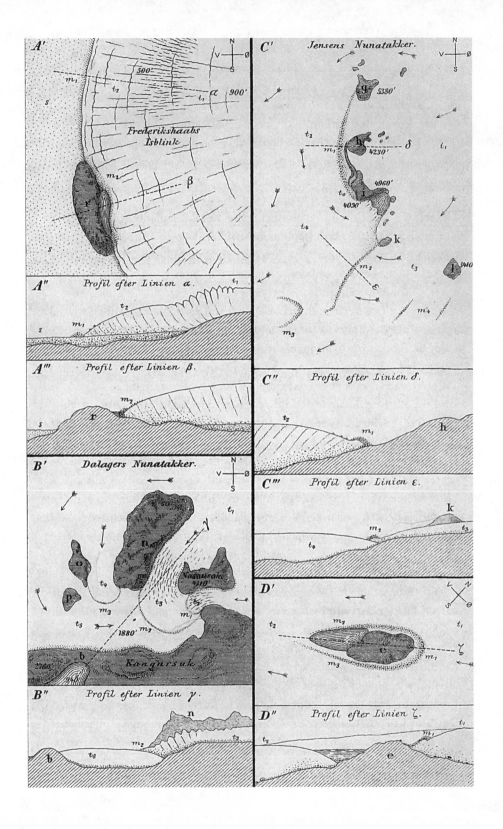

A'

300'

m_1

t_2

t_1 α 900'

s

Frederikshaabs Isblink

m_2 β

r

s

A'' Profil efter Linien α.

t_1

t_2

m_1

s

A''' Profil efter Linien β.

m_2

s r

B' Dalagers Nunatakker.

t_1

n

o γ

t_4 Nagssaio 3110'

p

m_3 t_3 m_1

t_5 1880' m_2

2760' b Kangursuk

950'

B'' Profil efter Linien γ.

n

m_2 t_3

b t_5

C' Jensens Nunatakker.

g 5380'

t_2 k δ t_1

m_1 4230'

4960'

t_0 i

4030'

t_4 k

m_2 t_3 l 5410

ε

m_3 m_4

C'' Profil efter Linien δ.

t_2 m_1 h

C''' Profil efter Linien ε.

k

m_2 t_3

t_4

D'

t_2 m_2 t_1

e ζ

m_3 m_1

D'' Profil efter Linien ζ.

t_2 m_1 t_1

e

◄ FIGURE 4.5. Chart illustrating the movements of the inland ice, showing the process by which the ice moved around and against the nunataks and how the pressure of the ice against the rock shaped the glaciers and created terminal moraines. Observing the nunataks was a way of studying the motion and behavior of very large bodies of glacial ice. Notice the nunataks named G, H, and I, which are also visible in Figure 4.6. *Source: Meddelelser om Grønland*, 1879, vol. 1, plate 5. Courtesy of the Scott Polar Research Institute.

▲ FIGURE 4.6. Illustration of travel over the ice in Greenland, showing the nunataks named G, H, and I, published in full color. *Source: Meddelelser om Grønland*, 1879, vol. 1, plate 3. Courtesy of the Scott Polar Research Institute.

scientific community permeated the pages of both journals. The expeditions in the 1870s had again shown that it was possible to carry out extensive surveying in the Arctic on a very tight budget, key reasons for the Danish government to commit early on to participating in the First IPY. However, there was another not-insignificant factor behind these efforts. The Danish territories in Greenland did not include the entire country. Surveying to determine its nature, which included what was under the ice sheet, was an important part of establishing imperial presence in the territory. That all of Greenland should be part of the Danish Kingdom was not a given, and some areas are still contested today. For example, Hans Island, named after Suersaq, is claimed by both Denmark and Canada. The significant increase in Danish expeditions to Greenland thereby reveals the interconnectedness and tensions between increased scientific internationalization on the one hand and nation building and imperial ambitions on the other.

The inland ice in Greenland was central to the development of the science of glaciology. In 1852 Rink published the first detailed reports on the character of Greenland's inland ice. The historian Tobias Krüger has argued that the concept of past ice ages implied that there had been huge ice sheets and glaciers, something many believed was improbable.[55] Rink's description of Greenland's interior as a vast ice sheet showed not only the possibility of such bodies, but also afforded the opportunity to study the phenomenon. In line with this, Kornerup wrote that in Greenland "you can, as no other place in the world, still today find the forces in action which in past times have shaped Scandinavia, Scotland, North America, and the Greenlandic coastland's ancient rocks."[56] Kane's travel narrative described these enormous glaciers, and it is interesting to note that Rink did not agree with Kane's descriptions of what he had named the Humboldt Glacier as something unique. Rather, this type of glacier could be seen all through the Greenland fjords. Under the coast of North Greenland there are places where strong currents keep the water from freezing in stream holes; Kane's assertion that they had seen an open water, the Open Polar Sea, Rink argued, was likely just such a stream hole. Rink first presented information that negated some of Kane's—an American—key findings to the English scientific scene in a talk at the Royal Geographical Society of London in 1858. One of those attending, the explorer and British naval officer Richard Collinson noted, "I think it very fortunate . . . that on this occasion we are acting the part of mediators, and not accusers, and that it has fallen to a Dane, and not an Englishman, to write this criticism." Indeed, Rink's detailed rejections of Kane's findings caused controversy or, as it was described by the American physicist Alexander Dallas Bache (1806–1867) in a follow-up article in the *Proceedings of the Royal Geographical Society*, a "feeling of vexation" in the American scientific circles.[57] Kane had recently passed away, which likely added to the strength of the reaction from the American Geographical Society.[58]

Rink took particular issue with the nature of Kane's narrative. Was it a scientific document or simply a description of a journey? He argued that while Kane described his narrative as an account "of the adventures of his fellow travelers" and "not a record of scientific investigations," it was "embellished with scientific theories extending far beyond the bounds of such a narrative."[59] Kane was engaging in scientific debates, Rink argued, therefore it was reasonable for him to address Kane's supposed discoveries and the theories behind them. Kane drew on the rhetorical strategy of framing his observations as being nothing but the direct observations of a traveler, as a way of distancing himself from charges of theoretical bias. In fact, he was highly attuned to the scientific milieu of the time and had prior to the second Grinnell expedition written and lectured on the theories supporting the Open Polar Sea. In Rink's opinion, Kane was presupposing the existence of the very thing he claimed Morton and Suersaq had found, and in doing so, falsely represented himself as an unbiased observer of natural phenomena.

In a move which further emphasizes the transnational nature of Arctic expeditions in the nineteenth century, Rink referred to Petersen's description of the second Grinnell expedition when evaluating Kane's supposed discovery of the Open Polar Sea, noting that Petersen was "a man well known to me," and his "communications bear the full impression of truth, and are written in a clear and simple style, without boasting and self-praise." His account "seems to give a clearer picture of its results than that which Kane has sketched." Kane, Rink argued, saw what he wanted to see and interpreted the observations of others as he wished. At a later meeting of the Royal Geographical Society of London, figures such as Collinson, Murchison, and Back attempted to smooth things over, and the American geographer and geologist John Henry Alexander (1812–1867) told the society that he was glad the whole issue had been settled. Alexander noted that he was "happy to see that, after the judicious and fair sifting which these observations of Dr. Kane have undergone, no greater error has been discovered, so that any of us, should we be inclined to transport ourselves to those inhospitable regions, may now rely upon being never out of our reckoning more than a few miles."[60] Rink, however, was not placated—international collaborations were rarely straightforward.

THE FIRST INTERNATIONAL POLAR YEAR

Nowhere is the tension between nation building and internationalization in the Arctic more evident than the debates surrounding the First IPY in 1882–1883. The IPY brought together researchers from multiple countries with the aim of undertaking systematic and coordinated scientific experiments and observations in the Arctic and Antarctic. Britain and Canada alone among the "old powers" in the Arctic did not initially commit to the venture and sent no representatives to the First International Polar Conference. As was argued in the *Standard*, the type of Arctic expedition proposed for the IPY was distinctly different from those previously organized by the British Royal Navy: "Why we have refrained from joining the other nations, it is needless discussing. Doubtless, the Admiralty have taken the best advice before declining to co-operate with them. Whatever their reasons are, we must remember that, though this work which they are about to undertake may be admirable from a theoretical point of view, it is not exploration. *C'est magnifique, mais ce n'est pas guerre* [*sic*], and in the Polar Basin war of the old sort is what the public expect for their money."[61] Linking wars to exploration in the Arctic in this way, was very apropos for the British Arctic experience: nothing quite said heroic Arctic exploration as venturing into the unknown and dying of scurvy along the way. By contrast, the IPY consisted of polar stations with predetermined (and already known) locations where researchers could focus on scientific objectives rather than loftier goals such as searches for the Northwest Passage and the Open Polar Sea. When compared to the nature of their previous expeditions, we can begin to see why the British (and Canadian) reaction to this new venture was less than lukewarm.

The IPY was initiated by the Austrian explorer Karl Weyprecht (1838–1881) and the German explorer Georg von Neumayer (1826–1909). Both Neumayer and Weyprecht believed that scientific activity in the Arctic would be advanced if measurements and observations were carried out simultaneously at different geographical locations. In Weyprecht's view, fieldwork should be systematic and cooperative, as opposed to the past exploratory expeditions. He fully explained his ideas in a presentation, "Fundamental Principles of Scientific Arctic Investigation" (*Programme des travaux d'une expedition polaire international*), delivered to the Academy of Sciences. The presentation was repeated at the forty-eighth meeting of German Naturalists and Physicians. It was also published as a pamphlet and translated into multiple languages; for example, the Danish newspaper, *Jyllandsposten*, published a long report on Weyprecht's presentation, which included a translation of Weyprecht's six principles for Arctic research.[62] The historian F. W. G. Baker has provided a more precise English translation:

1. Arctic exploration is of the greatest importance for a knowledge of the laws of nature.
2. Geographical discovery carried out in these regions has only a serious value inasmuch as it prepares the way for scientific exploration as such.
3. Detailed Arctic topography is of secondary importance.

4. For science in the Geographical Pole does not have a greater value than any other point situated in high latitudes.

5. If one ignores the latitude the greater the intensity of the phenomena to be studied the more favourable the place for an observational station

6. Isolated series of observations have only a relative value.[63]

For the venture to be successful, Weyprecht emphasized, it was imperative that all participants follow the same procedures and undertake the same observations in meteorology, magnetism, the auroras, and astronomy, in other words, that all participants adhere to a common Arctic science.

In 1874, the year prior to Weyprechts presentation, the Bremen Association for the German North Polar Passage established a committee that advocated an international collaboration in the Arctic with several observatory stations. Invitations were sent to Britain, Sweden, Norway, Russia, and the United States. This led to the First International Polar Conference in Hamburg in October 1879. It was not an easy task to organize an international collaborative effort of this scale. After the Second International Polar Conference took place in August 1880, only four countries committed to securing the funds required for participation: Denmark, Russia, Norway, and Austria. However, by May 1881 the Netherlands, the United States, and France also committed to setting up stations, followed by Sweden in June. At the Third International Polar Conference in August 1881, Britain was still notable by its absence. As Christopher Carter has argued, "Politics could have as much of an impact on the success of a scientific venture as the theories and techniques utilized during the venture."[64] Indeed, the First IPY was novel both in being the first large-scale enterprise of formal international scientific corporation, but also in the way the research program prioritized a different type of scientific practice compared to exploration ventures. In addition, the First IPY broke with past standards for publishing the Arctic experience.

While the magnetic crusade and the search for the lost Franklin expedition generated international collaborations, these were not intentionally or fully working together on an official level.[65] Expeditions to observe the transit of Venus in 1874 are another example of the increase in international scientific partnerships. However, nationalism continued to be a central stumbling block for any true international collaboration to take place in the nineteenth century.[66] The First IPY was fundamentally different from previous international joint efforts in the Arctic. Explorations throughout the 1870s and 1880s were characterized by

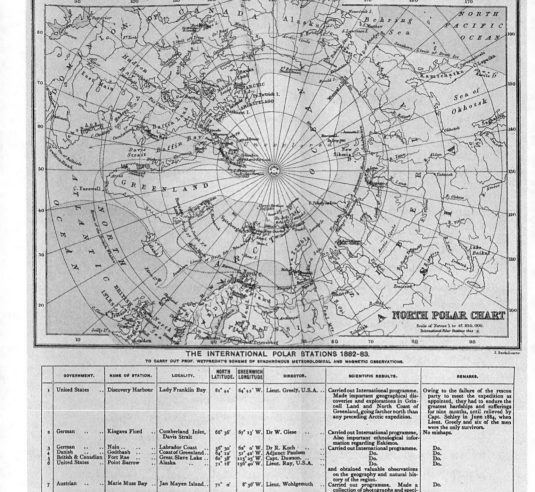

THE INTERNATIONAL POLAR STATIONS 1882-83.
TO CARRY OUT PROF. WEYPRECHT'S SCHEME OF SYNCHRONOUS METEOROLOGICAL AND MAGNETIC OBSERVATIONS.

	GOVERNMENT.	NAME OF STATION.	LOCALITY.	NORTH LATITUDE.	GREENWICH LONGITUDE.	DIRECTOR.	SCIENTIFIC RESULTS.	REMARKS.
1	United States ..	Discovery Harbour	Lady Franklin Bay	81° 44′	64° 45′ W.	Lieut. Greely, U.S.A. ..	Carried out International programme. Made important geographical discoveries and explorations in Grinnell Land and North Coast of Greenland, going farther north than any preceding Arctic expedition.	Owing to the failure of the rescue party to meet the expedition as appointed, they had to endure the greatest hardships and sufferings for nine months, until relieved by Capt. Sohley in June 1884, when Lieut. Greely and six of the men were the only survivors.
2	German ..	Kingava Fiord ..	Cumberland Inlet, Davis Strait	66° 36′	67° 13′ W.	Dr W. Giese ..	Carried out International programme. Also important ethnological information regarding Eskimos.	No mishaps.
3	German ..	Nain ..	Labrador Coast ..	56° 30′	62° 0′ W.	Dr R. Koch · ..	Carried out International programme.	Do.
4	Danish ..	Godthaab ..	Coast of Greenland..	64° 12′	51° 42′ W.	Adjunct Paulsen ..	Do.	Do.
5	British & Canadian	Fort Rae ..	Great Slave Lake ..	62° 38′	115° 25′ W.	Capt. Dawson. ..	Do.	Do.
6	United States ..	Point Barrow ..	Alaska ..	71° 18′	156° 40′ W.	Lieut. Ray, U.S.A. ..	Do. and obtained valuable observations on the geography and natural history of the region.	Do.
7	Austrian ..	Marie Muss Bay ..	Jan Mayen Island..	71° 0′	8° 36′ W.	Lieut. Wohlgemuth ..	Carried out programme. Made a collection of photographs and specimens of flora and fauna of district.	Do.
8	Swedish ..	Cape Thordsen ..	Spitzbergen ..	78° 30′	15° 30′ E.	Mr Eckholm	Carried out International programme.	Do.
10	Norwegian ..	Bossekop ..	Norway ..	69° 54′	26° 36′ E.	Mr Steen	Do.	Do.
11	Russian ..	Karmakuli ..	Moller Bay, Novaia Zemlia	72° 30′	53° 0′ E.	Lieut. Andreieff..	Do. Together with geographical research.	One death by accident.
12	Russian ..	Sagastir Island ..	Lena Delta ..	73° 0′	124° 42′ E.	Lieut. Jürgens ..	The first season's programme was successfully carried out.	The party volunteered for a second year's observations.
13	Dutch ..	Dickson Haven ..	Near Mouth of Yenisei	73° 55′	82° 0′ E.	Professor Snellen ..	Meteorological observations while beset in the ice at Waigat Strait.	The 'Varna' was beset in the ice at Waigat Strait, & the expedition did not reach its destination. They were rescued by the s.s.'Obi' in Sept.1883.

FIGURE 4.7. Map of polar stations in the First International Polar Year. Note that the stations are numbered 1–13 (with no number nine, which was cancelled after the stations had already been numbered), totaling twelve stations. *Scottish Geographical Magazine* 1, no 12 (1885): insert. Courtesy of the Scott Polar Research Institute.

national and imperial concerns about territorial control, in addition to increasing international scientific collaboration. In the end the IPY collaboration resulted in a total of fourteen expeditions, twelve of which were in the Arctic or sub-Arctic. Participating were Denmark, United States, Sweden, Russia, Norway, the Netherlands, Germany, France, the Austro-Hungarian Empire, Finland, Canada, and Britain. Three of the twelve polar stations in the Arctic were located within the Canadian Arctic: the German station by Kingua Fiord on Baffin Island, the American station in Lady Franklin Bay, and the British station at Fort Rae in the Northwest Territories. However, while Canada was invited to participate at the International Polar Conferences, there was no Canadian-organized polar station IPY. The Canadian involvement in the British station at Fort Rae was supportive but did not take part in determining the makeup of the expedition. Following Confederation in 1867—but not until 1880—Britain transferred the remaining islands in the High Arctic that were not already part of the Dominion to Canada.[67] As with the Danish claim to territorial ownership in Greenland, there were complications with making a stake for imperial governance in an area that had not yet been fully charted. Trevor Levere has pointed out that while the Canadian government (and the HBC) had made huge advances in mapping the country with its Geological Survey, the focus of the IPY was not geography but meteorology and geophysics. As the Royal Society of Canada was founded only in 1882, it could not lobby for a Canadian participation in the IPY the same way such societies had done in other countries.[68]

The potential value of a British contribution to the IPY, and the consequences if Britain sat out this venture, was discussed extensively in the periodical press. Weyprecht's criticism of the scientific achievements of past Arctic missions was met with a mixed response in Britain. An article in the *Times* remarked how Weyprecht was "convinced that the days of monster Arctic expeditions were past." It lamented the fact that the British government was reluctant to participate in the IPY, noting that "Weyprecht's scheme met with distinct approval everywhere, except among a few old-fashioned Arctic worthies in our own country, who were all for the fine old English method of expensive blundering." As was similarly stated in an article in the *Standard*, the scheme for the IPY would to "the impatient adventurers of the old school . . . sound sadly Academical, and tame to an unendurable degree." In contrast with the article in the *Times*, the *Standard* understood that such a venture with a focus on "pure science" was not "the work of the Admiralty" and would, perhaps regrettably, "do little to advance the naval renown of their respective countries."[69] *Geographical Magazine* countered

Weyprecht's notion that geographical discovery should not be the primary focus of Arctic expeditions: "Lieutenant Weyprecht complains of the prominence that has been given to geographical discovery in Arctic work, and that the conquest of physical difficulties has usurped the place of real scientific labour. As regards English scientific Arctic expeditions this complaint is groundless. Geographical discovery properly takes the first place, because it is by far the most important, and the conquest of physical difficulties is the means by which it is achieved."[70] Furthermore, the article scolded Weyprecht for including the Franklin search missions in his estimation of the scientific achievements of past British Arctic explorations, because these were not actual explorations. As discussed in the previous chapter, the aims and actual results of the search missions were difficult for many explorers and organizers to navigate. Search missions, the *Geographical Magazine* article contended, should be excluded in any estimation of the scientific value of past British Arctic explorations.

The criticism in the above-quoted passage nicely illustrates, the three interconnected reasons for the hesitant British response to the IPY. First, this was not a heroic exploratory Arctic expedition as indicated by the lack of "conquest of physical difficulties" associated with the polar stations. Second, if knowing equals owning (and geographical discovery had been a key way of stamping imperial authority on the Arctic), geography should be the primary objective. Third, the criticism of past Arctic explorations impugned all British ventures there. The experiences of Graah, Rae, and the HBC showed that the best methods for traveling while surviving in the Arctic were those developed and fine-tuned by the Indigenous peoples. The HBC had success with emulating these technologies, as did the explorers associated with the KGH, while the British Admiralty resisted changing their approach. As the proposed IPY was founded on the idea that international collaboration would achieve more than had been possible before, the implication was again that other nations were equal to, or better than, the British.

The British-Canadian station at Fort Rae contributed only the absolute minimum and produced a comparatively small amount of results. The scientific objectives of the IPY were divided into two groups, voluntary and obligatory. In addition, it was voluntary to make further observations, including in hydrography, atmospheric electricity, the nature and behavior of the ice, zoology, botany, and geology. The British-Canadian Fort Rae Polar Station was directed by a committee of the Royal Society of London, which consisted of Rae, George Richards (1820–1896), Robert Henry Scott (1833–1916), and George Stokes (1819–1903). Rae steadily lobbied the government and scientific societies to send out expeditions and also supported the British participation in the IPY. Markham also lobbied for a new government-organized expedition to the Arctic, but was staunchly opposed to the IPY, as he argued, "The discouragement of Arctic discovery for the sake of these impracticable observations is much to be deprecated."[71] Throughout the 1870s and early 1880s, Markham and Richards had recurring conflicts over the organization of Arctic expeditions. During Nares's 1875–1876 expedition, Markham wanted to send out a vessel to communicate with them, which Richards repeatedly opposed.[72] By 1880 their differences were even more strongly delineated. Markham noted that "Arctic discovery, wintering, and sledge exploration has an avowed enemy in Sir George Richards." This was a true assessment, as Richards thought would-be Arctic explorers used science as a "stalking horse" for their own ambitions. Echoing years of Franklin search missions as well as Nares's recent expedition, he argued that history showed that the organizers of exploratory expeditions were responsible for its safety, "which will probably mean a second one to be sent in search of it."[73]

In Markham's view, Richards was writing as someone who had no passion for "the good old cause" of Arctic exploration because of his own personal failure as an explorer. In a private letter Markham noted that Richards had "been over-rated" all his life, even though he was "a failure in the Arctic Regions," "a failure as a sledge traveler," "an utter failure as Hydrographer" and "brought the surveys down to the lowest ebb of inefficiency." Markham also criticized Rae and did not think much of his involvement. He noted that "another ill-wisher is Dr. Rae, but it does not much signify what he says. Richards, however, from the position he has held, might do real harm." By framing Richards as a failure in Arctic exploration, Markham was delegitimizing his authority on all matters pertaining to the polar regions. Behind the personal attacks, we see the crux of their disagreements, fundamentally to do with the design and purpose of Arctic explorations and the identity of the Arctic explorer. This was also reflected in the position of the Royal Geographical Society which opposed the IPY. Its position was that observing stations such as those proposed under the IPY should be considered as "subsidiary to the work of Geographical Exploration."[74]

In spite of such protests, and at the very last minute, Britain decided to contribute a polar station at Fort Rae by the Great Slave Lake.[75] The Canadian government supported the project with a small amount of money. The expedition party consisted of Captain Henry P. Dawson and C. S. Wedenby, both of the Royal Artillery, and Sergeant Instructor J. English and Sergeant F. W Cooksley, both of the Royal Horse Artillery. Their selection was in contrast to the previous Arctic expeditions organized by the British government, which had primarily consisted of men from the Royal Navy, due to the central difference between the Fort Rae Polar Station for the IPY and previous British expeditions. Weyprecht's vision for the IPY was based on the idea that deliberate and systematic observations could yield a more useful scientific output than what had been achieved from the exploratory expeditions. This was no longer opportunistic science dependent on the luck—or misfortune, as being frozen in for extended periods freed up time to undertake scientific observation—of the expeditions. It transformed the field into a laboratory in a much more institutionalized way than before. Or, as was described in the *Dublin Review*, "Stations are to be planted around the Pole in the form of a circle, at which simultaneous observations of the usual astronomical and meteorological elements are to be daily taken. And we are invited to believe that this will achieve a final and crowning victory to science. Such fancies could only enter the brain of an Arctic explorer." Many of the critical commentators did not think it was likely that this new version of Arctic science would be any better value for the money than previous ventures. The promise that the IPY would result in better understandings of the climate was the result of an "Arctic mania" and "the pet idea of polar explorers," who hoped to "bind and capture the Protean ice god and snatch from him the long-treasured secret."[76]

There is an interesting parallel between the move from exploratory Arctic expeditions to polar stations and the historical research on the relationship between the laboratory and the field. As Robert Kohler, among others, has pointed out, the lab–field border is more of a rhetorical and professional division that grew out of the so-called laboratory revolution between the 1840s and 1870s than an actual reflection of how researchers worked.[77] As there was a border, although intangible and constructed, between the laboratory and the field, there was also a marked division between scientific research at the polar stations and during the exploratory Arctic expeditions, which had geographical surveying as their main focus. The primary goal of the IPY was to produce internationally coordinated systematic meteorological and magnetical observations in the Arctic and Antarctic. International scientific cooperation was the hallmark of the IPY, as was the change of focus from geographical exploration to scientific observation. The IPY research was centered around the polar stations, which were largely sedentary, which had implications for the identity of the Arctic explorer-fieldworker. Many contemporary British commentators lamented that the inherently dangerous and heroic aspects of Arctic explorations were lost with the change of field site and the methodological transformations. The lack of enthusiasm for the IPY in Britain did not mean there was no interest in expeditions per se; rather, it was the character of this specific type of venture that was the issue.

The specialized focus of the IPY broke with the long tradition of British exploratory Arctic expeditions. When Ross in 1818 sailed in search of the Northwest Passage, his orders were first and foremost geographical and only second to "contribute to the advancement of science and natural knowledge."[78] The focus was broad and general, and as shown throughout this book, both the narratives and specialist scientific papers produced from the expeditions were utilized as evidentiary sources by researchers from many disciplines. Personal travel narratives were also important scientific documents and often included large supplements with detailed records of observations. This new venture was a land-based, science-oriented expedition that did not officially involve extensive geographical exploration, such as searches for the Northwest Passage or the North Pole. With a few exceptions, most of the expeditions under the IPY went according to plan. However, even in this more controlled field, disaster could still strike.

The United States contributed two stations, one in Point Barrow (Alaska had been purchased from Russia in 1867) and one in Lady Franklin Bay. Both American contributions consisted of crew members sent out by the US Army. The station at Point Barrow was small, consisting of only ten people. Point Barrow was, and is, home to several communities of Indigenous peoples, and the American IPY participants relied heavily on their assistance and support. Because of the relative ease and safety of the route to Point Barrow, and because of the contributions of local communities, this polar station was largely successful at controlling the field site as was intended. This was not the case for the Lady Franklin Bay expedition, led by Adolphus Greely (1844–1935). While the emphasis was no longer on geographical discovery, Greely and his crew first had to reach the chosen spot for the station. Lady Franklin Bay is located on the northeastern coast of what is now Nunavut, and his route was by ship, the *Proteus*, through Smith Sound. The expedition experienced difficulties in similar fashion to many of the past exploratory ventures, in particular those that also had traveled through Smith Sound. Out of the original twenty-five expedition members, nineteen died. Other participants also faced problems with their stations, though none to the same degree as Greely's crew. The Dutch expedition lost its ship and was stranded for ten months, and the Danish expedition was also stranded for an extended period.[79]

The research program of the First IPY therefore did not succeed in eliminating the types of disasters that so frequently followed the exploratory expeditions, in part because several of the IPY contributions involved voyaging through difficult waterways, such as Smith Sound. But for the expeditions that managed to safely reach their designated areas and were able to maintain their polar station through the support of the local communities, the new format for scientific research in the Arctic proved fruitful. For example, as with the American Point Barrow Polar Station, the British-Canadian contribution at Fort Rae was largely uneventful, and they were able to collect the data they had planned for.

The intended (though not always achieved) transformations in research style under the IPY were accompanied by changes in publication practices. The results from the Fort Rae Polar Station were not published as a traditional Arctic travel narrative, but appeared in several different forms; notably Dawson provided a brief preliminary "Report on the Circumpolar Expedition to Fort Rae" (1883) to the *Proceedings of the Royal Society of London* and later in full as *Observations of the International Polar Expeditions, 1882–83, Fort Rae* (1886). These were issued by the Royal Society. The prioritization of smaller-scale research expeditions under the IPY was very similar to the recent developments in Danish research in Greenland, reflected in *Meddelelserne*, the narrative style of which was also similar, with its combination of brief travel reports and scientific discussions.[80]

We also see the narrative results of the changes in scientific and exploratory practices in other ways. Seemingly none of the participants in the British-Canadian contribution published personal narratives after their journey. Dawson's *Observations* contained only a brief introduction that outlined their experiences at Fort Rae. By comparison, several accounts were written about the Lady Franklin Bay expedition, and these narratives reflect the more traditional exploratory nature of that venture. In addition to the formal publication of the IPY results, Greely himself published a narrative, *Three Years of Arctic Service: An Account of the Lady Franklin Bay Expedition of 1881–84, and the Attainment of the Farthest North*, in 1886. The book was dedicated to "its dead who suffered much—its living who suffered more." Comparing the official reports of the Lady Franklin Bay Polar Station to Greely's personal account illustrates the distinction made between the types of results and publications expected from the IPY and the standard personal narratives of a voyage of exploration. As Greely noted, *Three Years of Arctic Service* was based on his diary and was published "in response to the demands of the general public for a popular account" of his expedition.[81]

Both *Observations of the International Polar Expeditions* (1886) and

International Polar Expedition: Report on the Proceedings of the United States Expedition to Lady Franklin Bay, Grinnell Land (1888) provided detailed records of the expeditions' observations and experiments and were largely void of personal observations, aside from brief summaries of the expeditions.[82] *Observations* was divided into two sections: "Meteorological Observations" included Atmospheric Pressure, Air Temperature, Vapour Tensions and Relative Humidity, Wind, Amount Form and Direction of Clouds also Hydrometeors, Aurora, Solar Radiation, Terrestrial Radiation, Exposed Thermometer on Ground, and Earth Temperatures; and "Magnetical Observations" included Remarks (a summary), Declination, Horizontal Intensity, Vertical Intensity, Term Day Observations, Term Hour Observations, Selected Undisturbed Days, Selected Disturbed Days, and Journal of Auroral Observations. The primary focus at Fort Rae was meteorological and magnetic, the mandatory scientific observations of the IPY.

The instruments used at Fort Rae were all borrowed from the Kew Observatory, the Meteorological Office, and the Royal Geographical Society. The British government did not commit to participating in the IPY until April 1882 and consequently did not have time to custom-make or order new instruments. The majority of the expedition's food and other supplies were provided by the HBC in Winnipeg.[83] They departed from Liverpool on May 11, 1882, and reached Fort Rae on August 30 via Quebec, Winnipeg, and Carlton. It took them two months to travel from Carlton to Fort Rae. This was the roughest part of the journey, and some of the scientific instruments were at one point submerged underwater when they were hit by a gale. Luckily the instruments were not damaged, and the majority of the provisions were also salvaged. Despite the accident, "the performance of the magnetic instruments was satisfactory, with the exception of the balance magnetometer."[84] They began their meteorological observations the day after they arrived. Because Fort Rae was a preexisting establishment, they were able to convert a log hut used for storage into their magnetic observatory, which was finished in mid-September. Although they were stationary, they still encountered difficulties with their field site. Wild animals visited the station and disturbed the instruments. In an attempt to secure the site, particularly to keep out wolves, they decided to build a fence around their meteorological instruments during the winter.

Because the expedition was so small, one person carried out both the meteorological and magnetic observations. This was possible due to the proximity of the log-hut-turned-observatory. The expedition observed the aurora borealis, according to Weyprecht's systematic scheme, every night the sky was clear. They recorded the distribution in both local and Göttingen mean time, and the form and brightness of the aurora was evaluated according to a scale. As Weyprecht had determined that "isolated series of observations have only a relative value," scales such as these were utilized to enable a more standardized recording of observed phenomena.[85] The brightness of the aurora was indicated on a scale from one to four; interestingly it was also noted that on this scale five would be brighter than the Milky Way, and four, the actual maximum of the scale, would be bright enough to read by. They also noted the color of the aurora, viewed through the spectroscope, expressed by Roman figures corresponding to what it mostly resembled: arch, streamers, striæ, corona, patches or undefined light, dark segment, polar light, and sheaves. Readings of the magnetic instruments also followed a strict Weyprechtian system. Three readings were done with the same instrument, two taken one and a half minutes before and after the hour, and one at the hour. These and many other methodological choices and reflections were included in *Observations*, as were the instances when they encountered difficulties. Like the travel narratives from Arctic explorations before them, including such details added to the authority of the text. Perceived transparency and objectivity were central to establishing trust in their data, especially important in an international cooperative effort such as this.

In contrast with the optimism and enthusiasm that preceded Nares's expedition, the British response to the suggested IPY was surprisingly uninterested, as was the case in Canada, but not so in the United States and Denmark. For Canada and Britain, the expenditure associated with participation in a venture like the IPY was not easily justified. The style and objective of the proposed polar stations did not fit with the British trope of heroic Arctic explorations into the unknown. While Canada had plenty of experience in establishing stations in the Arctic through the HBC, and there had been a fruitful collaboration with American museums such as the Smithsonian in collecting natural history specimens, the primary motivation for Arctic expeditions had nearly always been geographical surveying, with other scientific goals occupying a secondary position. Therefore a focus on science relating to geophysics was a hard sell.

The polar stations were in the field, but "the field" was not a singular entity. Rather there were multiple types of fields constructed by those involved, be it supporters or critics. As Kohler argued, "The domains of laboratory and field are cultural domains first and foremost, where different languages, customs, material and moral economies, and ways of life prevail." The character of the field is therefore, to quote David Livingstone, "always politically negotiated" and "deeply uncontrollable."[86] When looking at Arctic science in this period, and the way it was discussed in the British periodical press, it is clear that the modifications to its setting, methods, and objectives of during the IPY also shifted the perceptions of what it meant to do fieldwork in the Arctic. The politics of fieldwork is reflected in the choice to send out men from the Royal Artillery instead of the navy. Even though the Royal Navy, together with the HBC, had dominated British and Canadian exploratory expeditions, the politics of the field site meant that the stationary, terrestrial Fort Rae could not be the venue of naval men. The methods of Arctic science also changed with the new field site. When science was a secondary priority, its results were largely determined by the individual preferences and skills of the crew. Attempts at standardizing fieldwork during the IPY were not completely new. As previous chapters have shown, the official instructions for exploratory Arctic expeditions often included highly detailed instructions for the preferred observations and experiments. Learned societies, private naturalists, and scientific instrument makers lent their expensive instruments to the missions, and the explorers dutifully recorded and compared their observations undertaken with instruments from different makers. Several of the officers even received additional scientific training prior to departing. Furthermore the HBC collaboration with the Smithsonian between the 1850s and 1870s also developed detailed instructions for collecting natural history specimens in an attempt to control fieldwork. Yet they could not regulate the field itself.[87] While the Arctic was still an unpredictable field site, the sedentary nature of the polar stations afforded a higher level of control.

The IPY marked a transformative event and was significant in furthering the networks between the international community of researchers. Scientifically the IPY was largely a success. As Weyprecht and Neumayer had predicted, the coordinated international program of observers carrying out systematized fieldwork in the Arctic generated extensive scientific results, which were published in multiple countries and provided evidentiary resources for years to come. As the president of the Royal Meteorological Society, John Knox Laughton, wrote in 1884, "The complete year's careful observations at such a station cannot but be exceedingly valuable."[88] It should be noted as an extraordinary peculiarity and testament to just how different this expedition was that the British-Canadian IPY participants have not been the subjects of biographies and little is known about their lives. At Fort Rae there was none of the traditional exploration drama, nothing to conquer, and there was nothing or no one to find. This was not a "heroic" Arctic exploration, and the four participants were not widely celebrated upon their return. In the First IPY we can see shifts in the relationship between identity making in exploratory and more sedentary fieldwork. When later explorers such as Robert Peary (1856–1920), Fridtjof Nansen, and Otto Nordenskjöld (1869–1929) again evoked heroic masculine ideals in their specific constructions of the explorer persona, they were much less reliant on performances of scientific expertise than their predecessors.[89] The First IPY did not cause a separation of geographical discovery and scientific fieldwork but was part of a wider shift in ideas in Europe and Euro-America about what could and should be prioritized by different types of Arctic fieldworkers.

While the British government had been reluctant to organize the Nares expedition to the North Pole exactly because of the cost in both monetary terms and human lives, it also did not greet the opportunity of the IPY in enthusiastic terms. The IPY did not instantaneously establish and secure international cooperation; both during the planning and after the event, nationalistic and imperial concerns influenced science in the Arctic. Global science, in all the concept's possible configurations, was made difficult to achieve in part because of significant geopolitical instability and competition between nations vying for control over imperial assets, as well as perceptions of the proper style and objectives of Arctic exploration. When the British prime minister Disraeli renewed government support for the British presence in the Arctic, it was a decision linked to other nations establishing themselves as powers there. The difference between the sentiments expressed prior to Nares's expedition and the lack of enthusiasm for the 1882 venture surely relates to the complete change in style of the explorations proposed under the IPY. Nationalism and imperialism were clear stumbling blocks for large-scale international collaborations. However, even the lukewarm participation of the British and Canadian governments showed what could be achieved when science and international collaboration, not geographical exploration and national concerns, were the main purpose for entering the icy north.

A NEW SCIENCE?

The period leading up to the first IPY was characterized by an increase in international collaborations, as well as a shift in imperial authority in the Arctic. After the disappearance of the last Franklin expedition, and the many search missions that followed, there was very little state support for new British expeditions. While Britain experienced Arctic fatigue, other nations such as Denmark and the

United States were increasing their presence in the polar region. The expeditions organized by the British government had largely followed the same blueprint since 1818, but as the previous chapters have illustrated, not all organizers were committed to the large two-vessel format. Such differences were pushed to the fore in the lead-up to the first IPY, with international cooperation as its hallmark. John Ambrose Fleming (1849–1945), who was one of the organizers of the Second IPY in 1932–1933, wrote about the first that "the immeasurable enhancement in the worth of polar observations through this coöperative endeavor has been amply demonstrated by the theoretical and practical applications of some twenty volumes of data obtained during that period."[90] However, international collaboration challenged perceptions of the identity of the explorer, the purpose and nature of Arctic exploration, and science conducted there.

One indication of the increased internationalization of Arctic science is Hinrich Rink's decision to publish Suersaq's memoir in English rather than Danish because he believed it would reach a larger audience that way. The early volumes of *Meddelelser om Grønland* show a similar trend. The journal was published in Danish, but included French abstracts as a way to make the knowledge available to a non-Danish audience. Later volumes were also translated into English. The publication of *Meddelelser om Grønland* further reflected the political ideology that the loss of territories could be compensated through an intensification of industrial and scientific efforts. There were key parallels between developments in Denmark and Canada at this point. There was a fruitful cooperation between the HBC and the Smithsonian Institution between 1855 and 1865. As an extension of the HBC's scientific network, the Smithsonian was prioritized over British scientists in part because of how they rewarded the collectors. The increased American interest in the Arctic is evident in the purchase of Alaska, as well as in the US support for Arctic ventures. Three of the four expeditions in search of the North Pole that ventured through Smith Sound were American. That the fourth, the George Nares expedition, was British is an example of how significant national pride was to the organization of Arctic explorations, especially in Britain. Meanwhile Suersaq participated in four expeditions that all went through Smith Sound, and his travel narrative was a memoir that reflected on and challenged the perceived authority of them all.

While the prospect of economic gains through the extraction of resources was a key factor behind the intensification of Danish explorations of Greenland and the publication of scientific knowledge about the region, Suersaq's memoir also reveals another shift; ideas about who was an authoritative observer of Arctic phenomena were changing. Suersaq was a cultural intermediary, and there was tension in the reviews of his narrative between accepting him as an authoritative writer while at the same time framing him as a "child of nature." As an Inuk, Suersaq's authority on exploratory matters did not fit comfortably with the established perceptions of the Arctic explorer. With scientific research in the Arctic becoming an increasingly global pursuit, the issue of the explorer's identity was being slowly redefined. This was particularly clear from the British-Canadian participation in the IPY, when many British commentators noted, and often decried, this shift and how it was linked to the identity of the explorer. Prior to the IPY, expeditions had been exploratory or rescue missions, while the IPY was centered on polar stations. The largely sedentary nature of these stations, and their lack of focus on geographical discovery, had important methodological implications for scientific research in the Arctic. The polar stations provided a more stable field site for scientific pursuits and, together with the international commitment to following a set framework of what it should accomplish, had a profound impact on Arctic science. While exploratory missions continued to play an important part, the IPY demonstrated what could be accomplished scientifically through deliberate international cooperation in the Arctic.

EPILOGUE

Scientific practice in the Arctic changed dramatically in the nineteenth century. There was a transition from the early, scattered collecting of knowledge about the natural world in the region to an attempt at a more unified Arctic science by the early 1880s. This shift also reflected a change in the types of output expected from Arctic explorers, as well as how expeditions were funded and organized. If, as Weyprecht suggested, those key scientific observations that previously took up so much of an explorer's time, particularly those relating to meteorology and geophysics, were relegated to a new type of specialist practitioner who spent extended periods of time in semipermanent polar stations, what became of the stereotypical nineteenth-century heroic Arctic explorer?

If I today venture to my local bookshop (or open my browser), I will be able to pick from an abundance of travel accounts. In them I might find descriptions of foreign people in faraway lands. They may loosely take the shape of a bildungsroman, or a more practical tour guide from the *Lonely Planet* series. No matter the geographical focus of the book, it is likely to promise a story of adventure and invite the reader along for the ride. These aspects of travel writing would be familiar to the nineteenth-century reader. I first encountered travel literature as a child, reading, without much discrimination, books that both explicitly framed themselves as fiction and those that claimed to be true accounts of a personal voyage. As a young consumer, I watched exhilarating accounts of space and the deep sea, read books such as *Gulliver's Travels* by Jonathan Swift (though most of the satire was lost on me) and Henrik Pontoppidan's *The Polar Bear: A Portrait*, all of which sparked my imagination with the wonders of travel. What I did not see then, and what I am unlikely to find in a modern travel account now, are the specific aspects that made a nineteenth-century travel narrative also a scientific text.

The notion that Arctic explorers had to contribute extensive results to science slowly waned, generally speaking, as the nineteenth century progressed. The British Arctic expedition led by Nares in 1875 which produced extensive scientific results was a key exception, one that showed the persistent adherence to specific tropes of British exploration and knowledge-making. The First International Polar Year played a leading role in this transition. Its organization reflects a division of labor in Arctic expedition. The explorer conceptualized as a European or North American man (or occasionally woman) who bravely put themselves in danger to reach and expand the limits of geographical knowledge was not the type of fieldworker required for scientific investigations under the IPY. Those in favor of the IPY argued that in order to truly achieve valuable scientific results in the Arctic, it was necessary to set aside ambitions of geographical discovery. In turn this reflected a change in the methods that were deemed best suited for geographical discovery. The style of Arctic exploration organized by the British government throughout the nineteenth century typically required the explorer to navigate the high expectations that came when a venture consisted of large and expensive vessels with an equally large and expensive collection of scientific instruments.

When the American explorer Robert Peary embarked on his multiple attempts to reach the North Pole at the turn of the twentieth century, his style of exploration was markedly different to that of, say, Franklin. This was the case both in the organization of his travels and in the activities he undertook along the way. For Peary the main objective of his Arctic travels was reaching the North Pole—which he claimed to have achieved during his 1908–1909 expedition.[1] While the primary objective of the British-, Danish-, and HBC-organized expeditions in the first half of the nineteenth century had been geographical surveying, their secondary goal was scientific advancement. Although the main focus changed with the disappearance of Franklin's last expedition, scientific research and geographical discovery remained key preoccupations. But Peary had different priorities, which was reflected in the accounts he published. Peary did still collect natural history and ethnographic data, and he did undertake magnetic observations, but the manner in which he did so was more similar to how Rae had worked than Richardson. Peary published and lectured extensively as a way to gather funds and support for further expeditions, as did his wife Josephine Diebitsch Peary, who accompanied him on parts of his expeditions. Robert and Josephine both described the Arctic in texts that were highly readable—and almost completely void of the types of detailed scientific observations and measurements that earlier narratives had included within their texts and appendixes. Peary's emphasis was on the process of geographical discovery, on reaching the North Pole—*Ultima Thule*.

What Robert and Josephine Peary did discover was that veracity was still as fleeting a quality as it had been for John Ross in 1818. When Peary returned to the United States from his 1908–1909 expedition, he was met with news that explorer Frederick Cook (1865–1940) claimed to have reached the North Pole before him. How did Peary and Cook each make their case? It turns out that scientific observations and self-portrayal in their narratives were still a central part of constructing a trustworthy account. The American journalist and explorer Walter Wellman (1858–1934) concluded in an article in the *New York Times*, that there were "three ways to test the good faith of one who claims to have been to the pole—first, by his character: second, by his narrative; third, by his astronomical observations." Astronomical observations were important because the extent to which the explorer judiciously recorded such data contributed to the overall reliability of his claim of having reached a certain geographical point—that point of course being determined through such observations. Yet, Wellman noted, if the character and narrative are of questionable veracity, the astronomical data cannot serve as evidence on their own, "for the simple reason that, having concocted a story, such a man would not hesitate to concoct astronomical observations to match it—something very easily done."[2]

It was never a given that a narrative on its own was accepted as a true account of the Arctic. Of course veracity was linked to the author, but the surrounding circumstances of the expedition, and the textual strategies employed in the narrative, were equally significant in the construction of truthfulness. This process was never stable, and varied at different points in time, as well as in place. Yet, as Wellman's breakdown of the evidentiary hierarchy illustrates, the "personal equations," as Kuklick called it, were a central aspect of narrating the Arctic.[3] Identity-making was the key to creating credibility. We see this across national contexts, although the expression of this personal trustworthiness varied. By examining perceptions of the explorers' identity and the often-conflicting interests of imperialism and internationalism from a transnational perspective, we can see the complex construction and practice of science in the field. The extremeness of the Arctic, with its intense isolation, harsh environment, and acute danger, highlights clearly how science is shaped by its location. As a field site the Arctic was inherently uncertain, and the metropole had very little control over the types of results generated from these ventures, as the level of commitment to scientific pursuits depended on the interests and abilities of the explorer-fieldworkers and the support of Indigenous peoples, as well as the luck of the expedition. Upon their return back to the metropole, it was also not straightforward for explorers and the organizing bodies to control how their venture was perceived. This shows the complexity and the multidirectional nature of scientific knowledge, which is not limited to the Arctic, but applies to field science across the globe more generally. What comes to the fore is the instability of nineteenth-century scientific practices and narrative strategies in the field, in the Arctic, and beyond.

NOTES

Introduction

Epigraph: John Barrow, *Voyages of Discovery and Research within the Arctic Regions, from the Year 1818 to the Present Time: Under the Command of the Several Naval Officers Employed by Sea and Land in Search of a Northwest Passage from the Atlantic to the Pacific; with Two Attempts to Reach the North Pole. Abridged and Arranged from the Official Narratives, with Occasional Remarks* (London: John Murray, 1846), v–vi.

1. John Barrow, *Voyages of Discovery*, 530, 11, 12.

2. John Barrow, *Voyages of Discovery*, vi.

3. James Fleming and Vladimir Jankovic, "Revisiting Klima," *Osiris* 26, no. 1 (2011): 4–6; Kirsten Hastrup, "Anticipating Nature: The Productive Uncertainty of Climate Models," in *The Social Life of Climate Change Models: Anticipating Nature*, ed. Kirsten Hastrup and Martin Skrydstrup (New York: Routledge, 2012), 14.

4. Katharine Anderson, *Predicting the Weather: Victorians and the Science of Meteorology* (Chicago: University of Chicago Press, 2005), 260. See also Richard H. Grove, *Green Imperialism: Colonial Expansion, Tropical Island Edens and the Origins of Environmentalism, 1600–1860* (Cambridge: Cambridge University Press, 1995), 9.

5. Balfour Stewart, "Arctic Exploration," *Times* [London], Dec. 21, 1872, 10.

6. "The British Association," *Standard*, Aug. 25, 1882.

7. "The Arctic Campaign," *Standard*, Apr. 14, 1882.

8. Michael Trevor Bravo, "The Postcolonial Arctic," *Moving Worlds: A Journal of Transcultural Writings* 15 (2015): 105.

9. Although some expeditions went farther north of permanent human inhabitation, they still relied on Indigenous knowledge and labor, and used the northernmost towns as their base for stocking up on supplies.

10. Mary Louise Pratt, *Imperial Eyes: Travel Writing and Transculturation* (London: Routledge, 1992), 7; Stuart B. Schwartz, ed., *Implicit Understandings: Observing, Reporting and Reflecting on the Encounters between Europeans and Other Peoples in the Early Modern Era* (Cambridge: Cambridge University Press, 1994). See also Tiffany Shellam, "Miago and the 'Great Northern Men': Indigenous Histories from In-Between," in *Indigenous Mobilities: Across and Beyond the Antipodes*, ed. Rachel Standfield, ANU Press, 2018, 185–208, www.jstor.org/stable/j.ctv301dn7.12.

11. I draw in particular on the insights regarding the role of economics for the history of science by Casper Andersen, Jakob Bek-Thomsen, and Peter C. Kjærgaard, "The Money Trail: A New Historiography for Networks, Patronage, and Scientific Careers," *Isis* 103, no. 2 (2012): 310–15.

12. Trevor H. Levere, *Science and the Canadian Arctic: A Century of Exploration, 1818–1918* (Cambridge: Cambridge University Press, 2004).

13. For a brief overview of Greenland's demography and economy, see Ole Marquardt, "Greenland's Demography, 1700–2000: The Interplay of Economic Activities and Religion," *Études/Inuit/*

Studies 26, no. 2 (2002): 47–69. Detailed histories of the HBC include Elle Andra-Warner, *Hudson's Bay Company Adventures: Tales of Canada's Fur Traders*, 2nd ed. (Victoria: Heritage House, 2003); Ted Binnema, *Enlightened Zeal: The Hudson's Bay Company and Scientific Networks, 1670–1870* (Buffalo, NY: University of Toronto Press, 2014).

14. See for example Marianne Rostgaard and Lotte Schou, *Kulturmøder i Dansk Kolonihistorie* (Copenhagen: Gyldendal Uddannelse, 2010), 21; Spencer Apollonio, *Lands That Hold One Spellbound: A Story of East Greenland* (Calgary: University of Calgary Press, 2008), 7–12; Peter A. Toft and Inge Høst Seiding, "Circumventing Colonial Policies: Consumption and Family Life as Social Practices in the Early Nineteenth-Century Disko Bay," in *Scandinavian Colonialism and the Rise of Modernity: Small Time Agents in a Global Arena*, ed. Magdalena Naum and Jonas M. Nordin (New York: Springer Science & Business Media, 2013), 107.

15. The key literature on the HBC includes Shepard Krech III, ed., *The Subarctic Fur Trade: Native Social and Economic Adaptations* (Vancouver: UBC Press, 2011); Andra-Warner, *Hudson's Bay Company Adventures*; Harold Adams Innis, *The Fur Trade in Canada: An Introduction to Canadian Economic History*, rev. ed. (Toronto: University of Toronto Press, 1999); Edith Burley, *Servants of the Honourable Company: Work, Discipline, and Conflict in the Hudson's Bay Company, 1770–1870* (Toronto: Oxford University Press, 1997); Arthur J. Ray and Donald B. Freeman, *"Give Us Good Measure": An Economic Analysis of Relations between the Indians and the Hudson's Bay Company before 1763* (Toronto: University of Toronto Press, 1978).

16. The best work on *Flora Danica* is Henning Knudsen, *Fortællingen om Flora Danica* (Copenhagen: Statens Naturhistoriske Museum, Lindhardt og Ringhof, 2014).

17. For a general overview of visuality and the representation of the Arctic, see Russell A. Potter, *Arctic Spectacles: The Frozen North in Visual Culture, 1818–1875* (Seattle: University of Washington Press, 2007); Eavan O'Dochartaigh, "The visual culture of the Franklin Search Expeditions to the Arctic (1848–55)" (PhD diss., National University of Ireland, Galway, 2018); I. S. MacLaren, "The Aesthetic Map of the North, 1845–1859," *Arctic* 38, no. 2 (June 1985): 89–103. For more on Arctic panoramas, see Jen Hill, *White Horizon: The Arctic in the Nineteenth-Century British Imagination* (Albany: State University of New York Press, 2009), 130–84; Russell A. Potter and Douglas W. Wamsley, "The Sublime yet Awful Grandeur: The Arctic Panoramas of Elisha Kent Kane," *Polar Record* 35, no. 194 (July 1999): 193–206; Ralph O'Connor, *The Earth on Show: Fossils and the Poetics of Popular Science, 1802–1856* (Chicago: University of Chicago Press, 2008), 269, 274. For more on science and photography, see Geoffrey Belknap, *From a Photograph: Authenticity, Science and the Periodical Press, 1870–1890* (London: Bloomsbury, 2016).

18. Daniela Bleichmar, *Visible Empire: Botanical Expeditions and Visual Culture in the Hispanic Enlightenment* (Chicago: University of Chicago Press, 2012), 8. See also Efram Sera-Shriar, "Arctic Observers: Richard King, Monogenism and the Historicisation of Inuit through Travel Narratives," *Studies in History and Philosophy of Science Part C: Studies in History and Philosophy of Biological and Biomedical Sciences* 51 (June 2015): 23–31.

19. Jane Burbank and Frederick Cooper, *Empires in World History: Power and the Politics of Difference* (Princeton, NJ: Princeton University Press, 2010).

20. Janet Browne, "Biogeography and Empire," in *Cultures of Natural History*, ed. Nicholas Jardine, James A. Secord, and E. C. Spary (Cambridge: Cambridge University Press, 1996), 306–14.

21. Henrika Kuklick, "Personal Equations: Reflections on the History of Fieldwork, with Special Reference to Sociocultural Anthropology," *Isis* 102, no. 1 (2011): 1–33; Innes M. Keighren, Charles W. J. Withers, and Bill Bell, *Travels into Print: Exploration, Writing, and Publishing with John Murray, 1773–1859* (Chicago: University of Chicago Press, 2015), 17.

22. James Clifford, *Routes: Travel and Translation in the Late Twentieth Century* (Cambridge, MA: Harvard University Press, 1999), 8. See also Shino Konishi, Maria Nugent, and Tiffany Shellam, *Indigenous Intermediaries: New Perspectives on Exploration Archives*, ANU Press, 2015, www.jstor.org/stable/j.ctt19705zg.

23. Hill, *White Horizon*; Robert G. David, *The Arctic in the British Imagination 1818–1914* (Manchester: Manchester University Press, 2000); Alexander Kraus, "Scientists and Heroes: International Arctic Cooperation at the End of the Nineteenth Century," *New Global Studies* 7, no. 2 (2013): 101–16; Bruce Hevly, "The Heroic Science of Glacier Motion," *Osiris* 11 (Jan. 1996): 66–86.

24. For discussions about the relationship between fieldwork and armchair theorizing, see Efram Sera-Shriar, *The Making of British Anthropology, 1813–1871* (London: Pickering and Chatto; Pittsburgh: University of Pittsburgh Press, 2013); Sera-Shriar, "Arctic Observers"; George W. Stocking, *The Ethnographer's Magic and Other Essays in the History of Anthropology* (Madison: University of Wisconsin Press, 1992); Stocking, *Observers Observed: Essays on Ethnographic Fieldwork* (Madison: University of Wisconsin Press, 1984); Kuklick, "Personal Equations"; Richard C. Powell, "Becoming a Geographical Scientist: Oral Histories of Arctic Fieldwork," *Transactions of the Institute of British Geographers*, n.s., 33, no. 4 (Oct. 2008): 548–65; Sera-Shriar, "What Is Armchair Anthropology? Observational Practices in 19th-Century British Human Sciences," *History of the Human Sciences* 27, no. 2 (2014): 26–40.

25. Elizabeth A. Bohls and Ian Duncan, eds., *Travel Writing 1700–1830: An Anthology*, Oxford World's Classics (Oxford: Oxford University Press, 2008), xx; Keighren, Withers, and Bell, *Travels into Print*, 7; Claire Jowitt and Carey Daniel, eds., *Richard Hakluyt and Travel Writing in Early Modern Europe* (Farnham, UK: Ashgate, 2012), 4.

26. Pratt, *Imperial Eyes*; Miguel A. Cabañas, Jeanne Dubino, Veronica Salles-Reese, and Gary Totten, "Introduction" in Cabañas, Dubino, Salles-Reese, and Totten, eds., *Politics, Identity, and Mobility in Travel Writing* (New York: Routledge, 2015), 1(quote).

27. "By Land and Sea," *Fraser's Magazine for Town and Country, 1830–1869; London*, Sept. 1853, 249.

28. Steven Shapin, *The Scientific Life: A Moral History of a Late Modern Vocation* (Chicago: University of Chicago Press, 2009), 6; Steven Shapin and Simon Schaffer, *Leviathan and the Air-Pump: Hobbes, Boyle, and the Experimental Life*, rev. ed. (Princeton, NJ: Princeton University Press, [1985] 2011), 60. For more on historiographical issues with truth judgments, see Keighren, Withers, and Bell, *Travels into Print*, 11.

29. Adriana Craciun, "Writing the Disaster: Franklin and Frankenstein," *Nineteenth-Century Literature* 65, no. 4 (Mar. 2011): 433–80; Hill, *White Horizon*; Janice Cavell, *Tracing the Connected Narrative: Arctic Exploration in British Print Culture, 1818–1860* (Toronto: University of Toronto Press, 2008).

30. In Britain the paper tax refers to a system of taxation on printed media, with the Stamp Act of 1712. The tax on newspapers was a way for the government to control the press, and a so-called stamp duty was enforced on newspapers until 1855. The paper duty was removed in 1861. For more on this topic, see Martin Hewitt, *The Dawn of the Cheap Press in Victorian Britain: The End of the "Taxes on Knowledge," 1849–1869* (London: Bloomsbury, 2014).

31. Geoffrey Cantor et al., *Science in the Nineteenth-Century Periodical: Reading the Magazine of Nature* (Cambridge: Cambridge University Press, 2004), 1. For further studies focusing on the British context for periodical publishing, see Jonathan R. Topham, "Beyond the 'Common Context': The Production and Reading of the Bridgewater Treatises," *Isis* 89, no. 2 (June 1998): 233–62; Topham, "Scientific Publishing and the Reading of Science in Nineteenth-Century Britain: A

Historiographical Survey and Guide to Sources," *Studies in History and Philosophy of Science Part A* 31, no. 4 (2000): 559–612; Bernard Lightman, *Victorian Popularizers of Science: Designing Nature for New Audiences* (Chicago: University of Chicago Press, 2009); Aileen Fyfe, *Steam-Powered Knowledge: William Chambers and the Business of Publishing, 1820–1860* (Chicago: University of Chicago Press, 2012); Aileen Fyfe, *Science and Salvation: Evangelical Popular Science Publishing in Victorian Britain* (Chicago: University of Chicago Press, 2004); Aileen Fyfe and Bernard Lightman, *Science in the Marketplace: Nineteenth-Century Sites and Experiences* (Chicago: University of Chicago Press, 2007); James A. Secord, *Victorian Sensation: The Extraordinary Publication, Reception, and Secret Authorship of Vestiges of the Natural History of Creation* (Chicago: University of Chicago Press, 2000); James A. Secord, "Knowledge in Transit," *Isis; an International Review Devoted to the History of Science and Its Cultural Influences* 95, no. 4 (Dec. 2004): 654–72.

32. Casper Andersen and Hans H. Hjermitslev, "Directing Public Interest: Danish Newspaper Science 1900–1903," *Centaurus* 51, no. 2 (May 2009): 143–67.

33. Ebbe Kühle, *Danmarks Historie i Et Globalt Perspektiv* (Copenhagen: Gyldendal, 2008), 178–85; Ole Feldbæk, *Danmarks Historie* (Copenhagen: Gyldendal, 2010), 185–301; Dan Christensen, *Hans Christian Ørsted: Reading Nature's Mind* (Oxford: Oxford University Press, 2013), 483–87; Robert Justin Goldstein, *Political Censorship of the Arts and the Press in Nineteenth-Century Europe* (New York: St. Martin's Press, 1989), 33.

34. George Fetherling, *The Rise of the Canadian Newspaper* (Toronto: Oxford University Press, 1990), preface; Merrill Distad, "Newspapers and Magazines," in *History of the Book in Canada: 1840–1918*, ed. Patricia Fleming, Yvan Lamonde, and Fiona Black (Toronto: University of Toronto Press, 2005), 2:293–302; Patricia Fleming, Yvan Lamonde, and Giles Gallichan, eds., *History of the Book in Canada: Beginnings to 1840*, 2 vols. (Toronto: University of Toronto Press, 2004–2005).

35. Michael Bravo and Sverker Sörlin, "Narrative and Practice—An Introduction," in Michael Bravo and Sverker Sörlin, eds., *Narrating the Arctic: A Cultural History of Nordic Scientific Practices*, 3–32 (Canton, MA: Science History Publications, 2002), 19. For a detailed overview on the theories of globalization, see Bruce Mazlish, "Comparing Global History to World History," *Journal of Interdisciplinary History* 28, no. 3 (1998): 385–95; Jürgen Osterhammel and Niels P. Petersson, *Globalization: A Short History*, trans. Dona Geyer (Princeton, NJ: Princeton University Press, 2005); Patrick Manning and Jerry H. Bentley, "The Problem of Interactions in World History," *American Historical Review* 101, no. 3 (1996): 771.

36. C. A. Bayly, Sven Beckert, Matthew Connelly, Isabel Hofmeyr, Wendy Kozol, and Patricia Seed, "On Transnational History," AHR Conversation, *American Historical Review* 111, no. 5 (Dec. 2006): 1441–64. For a detailed overview on the theories of globalization, see Mazlish, "Comparing Global History to World History"; Osterhammel and Petersson, *Globalization*; Manning and Bentley, "Problem of Interactions in World History."

37. Patricia Seed, as quoted in Bayly et al., "On Transnational History," 1458. See also, P. Horden and N. Purcell, "The Mediterranean and 'the New Thalassology,'" *American Historical Review* 111, no. 3 (June 2006): 739.

38. Sugata Bose, *A Hundred Horizons: The Indian Ocean in the Age of Global Empire* (Cambridge, MA: Harvard University Press, 2006), 4.

39. Oliver E. Williamson, *The Economic Institutions of Capitalism* (New York: Free Press, 1985), 30; Andrew Pickering, *Constructing Quarks: A Sociological History of Particle Physics* (Chicago: University of Chicago Press, 1999), 13.

Chapter 1: New Beginnings in the Arctic

Epigraph: Thomas Merton (pseud.), "Arctic Natural History," *Literary Magnet of the Belles Lettres, Science, and the Fine Arts, 1824–1826* 1, no. 1 (Jan. 1824): 51.

1. William Scoresby, *The Arctic Whaling Journals of William Scoresby the Younger: The Voyages of 1817, 1818 and 1820*, ed. C. Ian Jackson (New York: Routledge, 2009), 3:xxix; Trevor H. Levere, *Science and the Canadian Arctic: A Century of Exploration, 1818–1918* (Cambridge: Cambridge University Press, 2004), 40–41; Annette Watson, "William Scoresby," in *Encyclopedia of the Arctic*, ed. Mark Nuttall (New York: Routledge, 2012), 1850; Levere, *Science and the Canadian Arctic*, 41; Peter Fjagesund, *The Dream of the North: A Cultural History to 1920* (Amsterdam: Rodopi, 2014), 257.

2. Michael S. Reidy, *Tides of History: Ocean Science and Her Majesty's Navy* (Chicago: University of Chicago Press, 2009), 169; Levere, *Science and the Canadian Arctic*, 37.

3. Richard H. Grove, *Green Imperialism: Colonial Expansion, Tropical Island Edens and the Origins of Environmentalism, 1600–1860* (Cambridge: Cambridge University Press, 1995); Deborah Neill, *Networks in Tropical Medicine: Internationalism, Colonialism, and the Rise of a Medical Specialty, 1890–1930* (Stanford, CA: Stanford University Press, 2012); Nancy Leys Stepan, *Picturing Tropical Nature* (Ithaca, NY: Cornell University Press, 2001); Katharine Anderson, *Predicting the Weather: Victorians and the Science of Meteorology* (Chicago: University of Chicago Press, 2005). See also Robert G. David, *The Arctic in the British Imagination 1818–1914* (Manchester: Manchester University Press, 2000), xvi; Levere, *Science and the Canadian Arctic*, 41–44.

4. John Ross, *A Voyage of Discovery, Made under the Orders of the Admiralty, in His Majesty's Ships Isabella and Alexander, for the Purpose of Exploring Baffin's Bay, and Inquiring into the Probability of a North-West Passage* (London: John Murray, 1819).

5. Frederick William Beechey, *A Voyage of Discovery towards the North Pole: Performed in His Majesty's Ships Dorothea and Trent, under the Command of Captain David Buchan, R.N.; 1818; to Which Is Added, a Summary of All the Early Attempts to Reach the Pacific by Way of the Pole* (London: R. Bentley, 1843).

6. Christopher Lloyd, *Mr. Barrow of the Admiralty: A Life of Sir John Barrow* (London: Irvington Publishers, 1970). John Ross to J W [Groken], 1818, MS 999/7/1–6, Scott Polar Research Institute, University of Cambridge (hereafter SPRI).

7. Charles W. J. Withers and Innes M. Keighren, "Travels into Print: Authoring, Editing and Narratives of Travel and Exploration, c.1815—c.1857," *Transactions of the Institute of British Geographers*, n.s., 36, no. 4 (Oct. 2011): 6, 45.

8. Beechey, *Voyage of Discovery*, 9.

9. William J. H. Andrewes, ed., *The Quest for Longitude: The Proceedings of the Longitude Symposium, Harvard University, Cambridge, Massachusetts, November 4–6, 1993* (Cambridge, MA: Collection of Historical Scientific Instruments, Harvard University, 1996), 4–6; J. B. Hewson, *A History of the Practice of Navigation*, rev. ed. (Glasgow: Brown, Son & Ferguson, [1951] 1983), 226; Donald Launer, *Navigation through the Ages* (New York: Sheridan House, 2009), 3–6; James Edward McClellan III and Harold Dorn, *Science and Technology in World History: An Introduction*, rev. ed. (Baltimore: Johns Hopkins University Press, [1999] 2006), 268.

10. George W. Stocking, *Observers Observed: Essays on Ethnographic Fieldwork* (Madison: University of Wisconsin Press, 1984), 107. For more on the theme of objectivity and trust, see Theodore M. Porter, *Trust in Numbers: The Pursuit of Objectivity in Science and Public Life* (Princeton, NJ: Princeton University Press, 1995); Graeme Gooday, *The Morals of Measurement: Accuracy, Irony, and Trust in Late Victorian Electrical Practice* (Cambridge: Cambridge University Press, 2004).

11. Ross, *Voyage of Discovery*, ii.

12. Steven Shapin and Simon Schaffer, *Leviathan and the Air-Pump: Hobbes, Boyle, and the Experimental Life*, rev. ed. (Princeton, NJ: Princeton University Press, [1985] 2011), 60–62.

13. "Captain Ross, and Sir James Lancaster's Sound," ed. William Blackwood, *Blackwood's Edinburgh Magazine*, May 1819, 150.

14. Ross, *Voyage of Discovery*, 116–17.

15. Edward Sabine, *Remarks on the Account of the Late Voyage of Discovery to Baffin's Bay Published by J. Ross, R.N.* (London: R. and A. Taylor, 1819); John Ross, *An Explanation of Captain Sabine's Remarks on the Late Voyage of Discovery to Baffin's Bay* (London: John Murray, 1819)

16. Sabine, *Remarks on the Account*, 11; Ross, *Explanation*, 7–8.

17. James Clark Ross to John Ross, Apr. 13, 1819, MS 486/4/2, SPRI.

18. Janice Cavell, *Tracing the Connected Narrative: Arctic Exploration in British Print Culture, 1818–1860* (Toronto: University of Toronto Press, 2008), 67–74.

19. "Captain Ross's Voyage to Baffin's Bay," ed. William Jerdan, *Literary Gazette: A Weekly Journal of Literature, Science, and the Fine Arts*, Apr. 24, 1819, 261–63, 261.

20. "Polar Expedition," ed. Samuel Drew, *Imperial Magazine*, Aug. 1819, 702.

21. "Captain Ross's Voyage to Baffin's Bay," 263.

22. "ART. XIX.—A Voyage of Discovery Made under the Orders of the Admiralty in His Majesty's Ships Isabella and Alexander, for the Purpose of Exploring Baffin's Bay, and Inquiring into the Probability of a North-West Passage," ed. William Roberts, *British Review, and London Critical Journal, 1811–1825*, May 1819, 413–39, 419; "ART. VIII—A Voyage of Discovery, Made, under the Orders of the Admiralty, in H. M. Ships Isabella and Alexander, for the Purpose of Exploring Baffin's Bay, and Inquiring into the Probability of a North-West Passage," ed. William Chambers, *Edinburgh Monthly Review*, June 1819, 726–46, 736.

23. Alexi Baker, "Longitude Essays," Cambridge Digital Library, http://cudl.lib.cam.ac.uk/view/ES-LON-00023/1 (accessed Mar. 21, 2016).

24. For a detailed account of Barrow's life and work, see Fergus Fleming, *Barrow's Boys* (New York: Atlantic Monthly Press, 2000). Several major works on the HBC have been written; see especially Ted Binnema, *Enlightened Zeal: The Hudson's Bay Company and Scientific Networks, 1670–1870* (Buffalo, NY: University of Toronto Press, 2014); Edwin Ernest Rich, *Hudson's Bay Company 1670–1870*, vol. 1, *1821–1870*, 3 vols. (New York: Macmillian, 1961). The most thorough works on the HBC and the mapping of Canada are Richard I. Ruggles, *A Country So Interesting: The Hudson's Bay Company and Two Centuries of Mapping, 1670–1870* (Montreal: McGill-Queen's Press, 1991); Don W. Thomson, *Men and Meridians: The History of Surveying and Mapping in Canada*, vol. 3, *1966–69* (Ottawa: R. Duhamel, Queen's Printer, 1969).

25. Levere, *Science and the Canadian Arctic*, 191; Binnema, *Enlightened Zeal*; 130.

26. These were George Back, Robert Hood, John Richardson, John Hepburn, and Samuel Wilkes—although the latter was sent home early in the voyage.

27. John Franklin, *Narrative of a Journey to the Shores of the Polar Sea in the Years 1819, 20, 21 and 22, with an Appendix on Various Subjects Relating to Science and Natural History Illustrated by Numerous Plates and Maps* (London: John Murray, 1823), 4, 6, 149, 165, 279.

28. Keith J. Crowe, *A History of the Original Peoples of Northern Canada*, rev. ed. (Montreal: McGill-Queen's University Press, [1974] 1991), 79; June Helm, Teresa S. Carterette, and Nancy Oestreich Lurie, *The People of Denendeh: Ethnohistory of the Indians of Canada's Northwest Territories* (Iowa City: University of Iowa Press, 2000), 232–33; Crowe, *History of the Original Peoples*, 79; Harriet Gorham, "Tattannoeuck (Augustus)," *The Canadian Encyclopedia*, http://www.thecanadianencyclopedia.ca/en/article/augustus/ (accessed Jan. 31, 2017).

29. Franklin, *Narrative of a Journey*, 264.

30. Michael Trevor Bravo, "Ethnographic Navigation and the Geographical Gift," in *Geography and Enlightenment*, ed. David N. Livingstone and Charles W. J. Withers (Chicago: University of Chicago Press, 1999), 218.

31. Franklin, *Narrative of a Journey*, 225. By "slow travel," Akaitcho meant that they were inept, making the trip dangerous.

32. Franklin, *Narrative of a Journey*, 225.

33. Bravo, "Ethnographic Navigation," 203–4.

34. Franklin, *Narrative of a Journey*, 262, 475.

35. Franklin, *Narrative of a Journey*, 475.

36. Franklin, *Narrative of a Journey*, 480.

37. No known vital dates.

38. Anthony Brandt, *The Man Who Ate His Boots: Sir John Franklin and the Tragic History of the Northwest Passage* (New York: Random House, 2011), 89.

39. Franklin, *Narrative of a Journey*, 471.

40. John Franklin, *Narrative of a Second Expedition to the Shores of the Polar Sea, in the Year 1825, 1826 and 1827: Including an Account of the Progress of a Detachment to the Eastward by John Richardson; Illustrated by Numerous Plates and Maps. Published by Authority of the Right Honourable the Secretary of State for Colonial Affairs* (London: John Murray, 1828), 9–11; Helm, Carterette, and Lurie, *People of Denendeh*, 233.

41. Franklin, *Narrative of a Second Expedition*, 10.

42. Historian Catherine Lanone has argued that Akaitcho used the war as an excuse, a deflection to not engage with Franklin's party again, though this seems to underestimate the difficulties the T'atsaot'ine people were having. Catherine Lanone, "Arctic Romance under a Cloud: Franklin's Second Expedition by Land (1825–7)," in *Arctic Exploration in the Nineteenth Century: Discovering the Northwest Passage*, ed. Frédéric Regard (London: Pickering and Chatto; Pittsburgh: University of Pittsburgh Press, 2015), 106; I draw here on Richard Clarke Davis, *Lobsticks and Stone Cairns: Human Landmarks in the Arctic* (Calgary: University of Calgary Press, 1996), 144; Beryl Gillespie, "Yellowknife," in *Handbook of North American Indians: Subarctic*, ed. William C. Sturtevant (Washington, DC: Government Printing Office, 1978), 286.

43. Binnema, *Enlightened Zeal*, 146.

44. Franklin, *Narrative of a Journey*, xiii.

45. Franklin, *Narrative of a Journey*, 647; Levere, *Science and the Canadian Arctic*, 109–10.

46. John Richardson, ed., *Fauna Boreali-Americana, or, The Zoology of the Northern Parts of British America: Containing Descriptions of the Objects of Natural History Collected on the Late Northern Land Expeditions, under Command of Captain Sir John Franklin, R.N.*, 4 vols (London: John Murray, 1829–37), 1:frontispiece, x, ix, xi. See for example, William Jackson Hooker, ed., *Flora Boreali-Americana, or, The botany of the northern parts of British America, compiled principally from the plants collected by Dr. Richardson & Mr. Drummond on the late northern expeditions, under command of Captain Sir John Franklin, R.N. to which are added (by permission of the Horticultural Society of London) those of Mr. Douglas from north west America and other naturalists*, 2 vols. (London: Henry G. Bohn, [1829]–40); Thomas Drummond, "Sketch of a Journey to the Rocky Mountains and to the Columbia River in North America," in William Hooker, ed., *Botanical Miscellany* (London: John Murray, 1830), 1:178–219.

47. Richardson, *Fauna Boreali-Americana*, xviiii.

48. Franklin, *Narrative of a Journey*, 472–74.

49. Raymond E. Lindgren, *Norway-Sweden: Union, Disunion, and Scandinavian Integration* (Princeton, NJ: Princeton University Press, 1959), 8–10; Rasmus Glenthøj and Morten Nordhagen Ottosen, *Experiences of War and Nationality in Denmark and Norway, 1807–1815* (London: Palgrave Macmillan UK, 2014), 257–78; Shelagh D. Grant, *Polar Imperative: A History of Arctic Sovereignty in North America* (Vancouver: Douglas & McIntyre, 2010), 97–98.

50. Wilhelm August Graah, *Undersøgelses-Reise til Østkysten af Grønland: Efter Kongelign Befaling Udført i Aarene 1828–31* (Copenhagen: J. D. Qvist, 1832). For quotations I make use of the English translation, Wilhelm August Graah, *Narrative of an Expedition to the East Coast of Greenland, Sent by Order of the King of Denmark, in Search of the Lost Colonies*, trans. G. Gordon Macdougall, first English ed. (London: Royal Geographical Society of London, 1837).

51. Travel narratives were significant in shaping ethnographic knowledge, as examined by scholars such as Efram Sera-Shriar, "Arctic Observers: Richard King, Monogenism and the Historicisation of Inuit through Travel Narratives," *Studies in History and Philosophy of Science Part C: Studies in History and Philosophy of Biological and Biomedical Sciences* 51 (June 2015): 23–31; Catherine Hall, *Civilising Subjects: Metropole and Colony in the English Imagination 1830–1867* (Chicago: University of Chicago Press, 2002); Michael Bravo and Sverker Sörlin, eds., *Narrating the Arctic: A Cultural History of Nordic Scientific Practices* (Canton, MA: Science History Publications, 2002). For an overview of the historicization of humans in the nineteenth century, see Efram Sera-Shriar, ed., *Historicizing Humans: Deep Time, Evolution, and Race in Nineteenth-Century British Sciences* (Pittsburgh: University of Pittsburgh Press, 2018); Mark Bevir, ed., *Historicism and the Human Sciences in Victorian Britain* (Cambridge: Cambridge University Press, 2017).

52. William Scoresby, *Journal of a Voyage to the Northern Whale-Fishery: Including Researches and Discoveries on the Eastern Coast of West Greenland, Made in the Summer of 1822, in the Ship Baffin of Liverpool* (Edinburgh: Archibald Constable, 1823).

53. Graah, *Narrative of an Expedition*, 13.

54. Graah, *Narrative of an Expedition*, 160.

55. Dates unknown.

56. Graah, *Narrative of an Expedition*, 32.

57. Graah, *Narrative of an Expedition*, xiv.

58. Arnold Arboretum, *Sargentia: A Continuation of the Contributions from the Arnold Arboretum of Harvard University* (Cambridge, MA: Arnold Arboretum of Harvard University, 1943); 34. The most detailed account of *Flora Danica* and the collection strategies of Danish botany is Henning Knudsen, *Fortællingen om Flora Danica* (Copenhagen: Statens Naturhistoriske Museum; Lindhardt og Ringhof, 2014). See also *Flora Danica*, Det Kongelige Bibliotek, http://www.kb.dk/da/materialer/kulturarv/institutioner/DetKongeligeBibliotek/Billeder_oversigt/flora_danica.html (accessed Dec. 11, 2015).

59. Eric Hultén, *Flora of Alaska and Neighboring Territories: A Manual of the Vascular Plants* (Stanford, CA: Stanford University Press, 1968), 982; Mary Louise Pratt, *Imperial Eyes: Travel Writing and Transculturation* (London: Routledge, 1992), 111–14; Lisbet Koerner, *Linnaeus: Nature and Nation* (Cambridge, MA: Harvard University Press, 2009), 1; Martin J. S. Rudwick, *The Meaning of Fossils: Episodes in the History of Palaeontology* (Chicago: University of Chicago Press, 1976), 208.

60. Graah, *Narrative of an Expedition*, 49.

61. Graah, *Narrative of an Expedition*, 52.

62. Graah, *Narrative of an Expedition*, 52.

63. Graah, *Narrative of an Expedition*, 52–53.

64. Daniela Bleichmar, *Visible Empire: Botanical Expeditions and Visual Culture in the Hispanic En-*

lightenment (Chicago: University of Chicago Press, 2012); Sujit Sivasundaram, "Towards a Critical History of Connection: The Port of Colombo, the Geographical 'Circuit,' and the Visual Politics of New Imperialism, ca. 1880–1914," *Comparative Studies in Society and History* 59, no. 2 (Apr. 2017): 382.

65. Graah, *Narrative of an Expedition*, preface.

66. Sivasundaram, "Towards a Critical History of Connection."

67. Martin Kemp, *Seen/Unseen: The Visual Ideas behind Art and Science* (Oxford: Oxford University Press, 2006), 191; Lorraine Daston and Peter Galison, *Objectivit* (New York: Zone Books, 2007), 69–70.

68. This refers to the Danish folk high schools, founded on the ideals of the clergyman Nikolai F. S. Grundtvig (1783–1872). For more on science and the Danish Folkehøjskoler, see Hans Henrik Hjermitslev, "Naturvidenskabens Rolle På de Danske Folkehøjskoler," in Fay L. Nilsson and Anders Nilsson, eds., *Två Sidor af Samma Mynt? Folkbilding och Yrkesutbildning Vid de Nordiska Folkhögskolorna* (Lund: Nordic Academic Press, 2010), 111–38.

69. See for example, Hans Egede, *Det Gamle Grønlands nye Perlustration: Eller Naturel-Historie, og Beskrivelse over det Gamle Grønlands Situation, Luft, Temperament og Beskaffenhed* (Copenhagen: J. C. Groth, 1741).

70. Graah, *Narrative of an Expedition*, 119, 122–23.

71. Graah, *Narrative of an Expedition*, 116.

72. Graah, *Undersøgelses-Reise*, 114.

73. Anne McClintock, *Imperial Leather: Race, Gender, and Sexuality in the Colonial Contest* (New York: Routledge, 1995), 22.

74. Graah, *Narrative of an Expedition*, 117, 124.

75. Klaus Georg Hansen, "Wilhelm August Graah," in *Encyclopedia of the Arctic*, ed. Mark Nuttall (New York: Routledge, 2012), 763.

76. C. Pingel, "XXIX—W.A. Graah, Undersøgelsesreise til Østkysten af Grønland," *Maanedsskrift for Litteratur* 10 (1833), 647.

77. Hinrich [Henry] Rink, *Naturhistoriske Bidrag til en Beskrivelse af Grønland* (Copenhagen: L. Kleins Bogtrykkeri, 1857). For an example of the usage of Vahl's data, see "Ichyologiske Bidrag til den Grönlandske Fauna," *Det Kongelige Danske Videnskabernes Selskabs Skrifter: Naturvidenskabelig og Mathematisk Afdeling*, 7 (Copenhagen: Bianco Lunos, 1838), 93.

78. Binnema, *Enlightened Zeal*, 140; Levere, *Science and the Canadian Arctic*, 111–12.

Chapter 2: Financial Opportunities in the Arctic
Epigraph: "The Failure of a Fourth Attempt within These Seven Years, at the Discovery of a North West Passage," *Times*, Oct. 19, 1825, 2.

1. "The Failure of a Fourth Attempt," 2.

2. [John Barrow], "ART. I.—1. Narrative of a Second Voyage in Search of a Northwest Passage, and of a Residence in the Arctic Regions, during the Years 1829–30–31–32–33," *Quarterly Review*, July 1835, 5.

3. Maurice James Ross, *Polar Pioneers: John Ross and James Clark Ross* (Montreal: McGill-Queen's Press, 1994), 112.

4. John Ross, *A Treatise on Navigation by Steam: Comprising a History of the Steam Engine, and an Essay towards a System of the Naval Tactics Peculiar to Steam Navigation, as Applicable Both to Commerce and Maritime Warfare; Including a Comparison of Its Advantages as Related to Other Systems in the Circumstances of Speed, Safety and Economy, but More Particularly in That of the National Defence* (London: Longman, Rees, Orme, Brown, and Green, 1828), dedication page.

5. Daniel R. Headrick, *Power over Peoples: Technology, Environments, and Western Imperialism, 1400 to the Present* (Princeton, NJ: Princeton University Press, 2012), 187, 202; Ross, *Polar Pioneers*, 113.

6. John Ross and James Clark Ross, *Narrative of a Second Voyage in Search of a North-West Passage, and of a Residence in the Arctic Regions during the Years 1829, 1830, 1831, 1832, 1833* (London: A. W. Webster, 1835), 2.

7. The last name is given both as Ericsson and Erickson depending on the source. See also Olav Thulesius, *The Man Who Made the Monitor: A Biography of John Ericsson, Naval Engineer* (Jefferson, NC: McFarland, 2007).

8. Ross and Ross, *Narrative of a Second Voyage*, 10–11.

9. Ross, *Polar Pioneers*, 188.

10. John Braithwaite, *Supplement to Captain Sir John Ross's Narrative of a Second Voyage in the Victory, in Search of a North-West Passage. Containing the Suppressed Facts Necessary to a Proper Understanding of the Causes of the Failure of the Steam Machinery of the Victory, and a Just Appreciation of Captain Sir John Ross's Character as an Officer and a Man of Science* (London: Chapman & Hall, 1835).

11. Ross and Ross, *Narrative of a Second Voyage*, 6–7, 10.

12. Braithwaite, *Supplement to Captain Sir John Ross's Narrative*, 2–3.

13. John Ross, *Explanation and Answer to Mr. John Braithwaite's Supplement to Captain Sir John Ross's Narrative of a Second Voyage in the Victory, in Search of a Northwest Passage* (London: A. W. Webster, 1835), 2.

14. Ross and Ross, *Narrative of a Second Voyage*, 121.

15. Ross and Ross, *Narrative of a Second Voyage*, 124.

16. Ross, *Treatise on Navigation by Steam*, 4–5, 8.

17. Braithwaite, *Supplement to Captain Sir John Ross's Narrative*, 18.

18. Braithwaite, *Supplement to Captain Sir John Ross's Narrative*, 565; "Supplement to Captain Sir John Ross's Narrative of a Second Voyage in the Victory, in Search of a North-West Passage, Containing the Suppressed Facts Necessary to a Proper Understanding of the Causes of the Failure of the Steam Machinery of the Victory, and a Just Appreciation of Captain Sir John Ross's Character as an Officer and a Man," *Literary Gazette: A Weekly Journal of Literature, Science, and the Fine Arts*, Nov. 7, 1835, 712; "A Supplement to Captain Sir John Ross's Narrative of the Second Voyage in the Victory, in search of a North-West Passage; containing the Suppressed Facts necessary to a proper Understanding of the Causes of the Failure of the Steam Machinery of the Victory, &c. &c.," *Monthly Magazine*, Dec. 1835, 565.

19. Ross, *Polar Pioneers*, 194. Panoramas were a key way through which expeditions were presented to a broad audience. A panorama, *A View of the Continent of Boothia, Discovered by Captain Ross*, was exhibited at Leicester Square to celebrate Ross's expedition. This was not done for Ross's first voyage in 1818, though it had been for Buchan's part of the twin expedition.

20. [Barrow], "ART. I.-1. Narrative of a Second Voyage," 4, 30, 32.

21. "Narrative of a Second Voyage in Search of a North-West Passage, and of a Residence in the Arctic Regions, during the Years 1829, 1830, 1831, 1832, 1833, by Sir John Ross, C.B., K.S.A., K.C.S., &c. &c.," *Literary Gazette: A Weekly Journal of Literature, Science, and the Fine Arts*, May 9, 1835, 289, 290; "Ross's Expedition," *Chambers's Edinburgh Journal*, Feb. 1832–Dec. 1853, Nov. 28, 1835, 347.

22. [David Brewster], "ART. VII.—Narrative of a Second Voyage in Search of a Northwest Passage, and of a Residence in the Arctic Regions during the Years 1829, 1830, 1831, 1832, 1833," *Edinburgh Review, 1802–1929*, July 1835, 421; Janice Cavell, *Tracing the Connected Narrative: Arctic Exploration in British Print Culture, 1818–1860* (Toronto: University of Toronto Press, 2008), 51.

23. [Barrow], "ART. I.-1. Narrative of a Second Voyage," 23.

24. [Barrow], "ART. I.-1. Narrative of a Second Voyage," 23–25.

25. [Barrow], "ART. I.-1. Narrative of a Second Voyage," 38.

26. Sujit Sivasundaram, "Natural History Spiritualized: Civilizing Islanders, Cultivating Breadfruit, and Collecting Souls," *History of Science* 39, no. 4 (Dec. 2001): 417–43; Sivasundaram, *Nature and the Godly Empire: Science and Evangelical Mission in the Pacific, 1795–1850* (Cambridge: Cambridge University Press, 2005).

27. Anonymous Missionary, "Udtog af en Dansk Dames Dagbog, Ført i Grønland 1837–1838," 2 parts, *Læsefrugter,* Jan. and Feb. 1839. I am thankful to Matthias Kaalund Keller for providing me with a copy of this narrative. Johan Christian Wilhelm Funch, *Syv Aar i Nordgrönland* (Viborg, Denmark: Rabell, 1840). In this book, all translations from the Danish are mine unless otherwise noted.

28. Michael Mann, "'Torchbearers upon the Path of Progress': Britain's Ideology of a 'Moral and Material Progress' in India; An Introductory Essay," in Harald Fischer-Tiné and Michael Mann, eds., *Colonialism as Civilizing Mission: Cultural Ideology in British India,* 1–26 (London: Anthem Press, 2004), 4. See also Winfried Baumgart, *Imperialism: The Idea and Reality of British and French Colonial Expansion* (Oxford: Oxford University Press, 1982), 16; Catherine Hall, *Civilising Subjects: Metropole and Colony in the English Imagination 1830–1867* (Chicago: University of Chicago Press, 2002), 21.

29. See, for example, Hall, *Civilising Subjects;* Keith E. Yandell, *Faith and Narrative* (Oxford: Oxford University Press, 2001); A. Twells, *The Civilising Mission and the English Middle Class, 1792–1850: The "Heathen" at Home and Overseas* (Dordrecht, Netherlands: Springer, 2008); Gavin Murray-Miller, *The Cult of the Modern: Trans-Mediterranean France and the Construction of French Modernity* (Lincoln: University of Nebraska Press, 2017); Peter Fibiger Bang and Dariusz Kolodziejczyk, *Universal Empire: A Comparative Approach to Imperial Culture and Representation in Eurasian History* (Cambridge: Cambridge University Press, 2012).

30. Funch, *Syv Aar i Nordgrönland,* 52.

31. *Naturbeskaffenhed* is an awkward term that approximately translates to "quality of nature," similar in scope to natural history. Funch, *Syv Aar i Nordgrönland,* 53–56.

32. Funch, *Syv Aar i Nordgrönland;* "Døde," Den til Forsendelse med de Kongelige Brevposter Privilegerede Berlingske Politiske og Avertissementstidende, Mar. 12, 1867, 7, Statsbiblioteket, Aarhus Universitet; Thomas Hansen Erslew, *Almindeligt Forfatter-Lexicon for Kongeriget Danmark med Tilhørende Bilande, fra 1814 til 1840* (Copenhagen: Forlagsforeningens Forlag, 1843), 474–75; Selskabet for Danmarks Kirkehistorie, *Kirkehistoriske Samlinger* (Copenhagen: Akademisk Forlag, 1911), 51.

33. I am particularly informed by Bernard Lightman, *Victorian Popularizers of Science: Designing Nature for New Audiences* (Chicago: University of Chicago Press, 2009); Aileen Fyfe and Bernard Lightman, *Science in the Marketplace: Nineteenth-Century Sites and Experiences* (Chicago: University of Chicago Press, 2007); Aileen Fyfe, *Science and Salvation: Evangelical Popular Science Publishing in Victorian Britain* (Chicago: University of Chicago Press, 2004); Jonathan R. Topham, "Beyond the 'Common Context': The Production and Reading of the Bridgewater Treatises," *Isis* 89, no. 2 (June 1998): 233–62.

34. J[ohannes] Reinhardt, "Ichyologiske Bidrag til den Grönlandske Fauna," *Det Kongelige Danske Videnskabernes Selskabs Skrifter. Naturvidenskabelig og Mathematisk Afdeling,* no. 7 (1838): 83–196.

35. The original Danish is "i den bedste mening." Tine Bryld, *I den Bedste Mening* (Copenhagen: Gyldendal, 2010).

36. For the Danish context I draw especially on Karen Vallgårda, *Imperial Childhoods and Christian Mission: Education and Emotions in South India and Denmark* (Basingstoke, UK: Palgrave Macmillan, 2014). For the Canadian context, see Marina Morrow, Olena Hankivsky, and Colleen Varcoe, eds., *Women's Health in Canada: Critical Perspectives on Theory and Policy* (Toronto: University of To-

ronto Press, 2008); Alvyn Austin and Jamie S. Scott, eds., *Canadian Missionaries, Indigenous Peoples: Representing Religion at Home and Abroad* (Toronto: University of Toronto Press, 2005); Truth and Reconciliation Commission of Canada, *Canada's Residential Schools: The Métis Experience; The Final Report of the Truth and Reconciliation Commission of Canada*, McGill-Queen's Native and Northern Series 83 (Montreal: McGill-Queen's University Press, 2016). Another comparison is the so-called home children of early 20th-century Canada. These were young people from disadvantaged backgrounds who were sent to Canada to work on farms. See, for example, Phyllis Harrison, *The Home Children: Their Personal Stories* (Winnipeg: Watson and Dwyer, 1979); Daniel Gorman, *Imperial Citizenship: Empire and the Question of Belonging* (Manchester: Manchester University Press, 2010), 186.

37. Mary Louise Pratt, *Imperial Eyes: Travel Writing and Transculturation* (London, New York: Routledge, 1992), 106; Sherrill E. Grace, "Gendering Northern Narrative," in *Echoing Silence: Essays on Arctic Narrative*, ed. John George Moss, Canadian Electronic Library, Books Collection, Re-Appraisals, Canadian Writers 20 (Ottawa: University of Ottawa Press, 1997), 166.

38. Lightman, *Victorian Popularizers of Science*, 96–97.

39. Steffen Auring, *Dansk Litteraturhistorie*, vol. 5, *Borgerlig Enhedskultur 1807–48*, Dansk Litteraturhistorie (Copenhagen: Gyldendal, 1984), 403.

40. Casper Andersen and Hans H. Hjermitslev, "Directing Public Interest: Danish Newspaper Science 1900–1903," *Centaurus* 51, no. 2 (May 2009): 144.

41. Anonymous Missionary, "Udtog af en Dansk Dames Dagbog," part 1, 105, 106.

42. Colin Coates, "Like 'The Thames towards Putney': The Appropriation of Landscape in Lower Canada," *Canadian Historical Review* 74, no. 3 (Sept. 1993): 317–43.

43. Funch, *Syv Aar i Nordgrönland*, 37–38.

44. Funch, *Syv Aar i Nordgrönland*, 38, 7.

45. Funch, *Syv Aar i Nordgrönland*, 51.

46. Funch, *Syv Aar i Nordgrönland*, 54.

47. Anonymous Missionary, "Udtog af en Dansk Dames Dagbog," part 2, 231.

48. "Nyheder Fra Udlandet," *Den til Forsendelse med Brevposterne Kongelig Allernaadigst (Alene) Privilegerede Aarhuus Stifts-Tidende*, Apr. 24, 1838, 1, Statsbiblioteket, Aarhus Universitet.

49. Thomas Simpson, *Narrative of the Discoveries on the North Coast of America: Effected by the Officers of the Hudson's Bay Company during the Years 1836–39* (London: R. Bentley, 1843).

50. William Barr, ed., *From Barrow to Boothia: The Arctic Journal of Chief Factor Peter Warren Dease, 1836–1839* (Montreal: McGill-Queen's University Press, 2002), 7; David A. Armour, "Biography—Dease, John," *Dictionary of Canadian Biography*, vol. 5, *1801–1820*, http://www.biographi.ca/en/bio/dease_john_5E.html (accessed Nov. 2, 2016); Alexander Simpson, *The Life and Travels of Thomas Simpson: The Arctic Discoverer* (London: R. Bentley, 1845), 19–20; Ted Binnema, *Enlightened Zeal: The Hudson's Bay Company and Scientific Networks, 1670–1870* (Buffalo, NY: University of Toronto Press, 2014), 147.

51. William Richard Hamilton (1777–1859) was the presiding president of the society, and he awarded the medals, as recorded in Royal Geographical Society of Great Britain, "The President's Address on Presenting Medals," *Journal of the Royal Geographical Society* 9 (1839): xi; "Arctic Expeditions: The Late Mr. Simpson," *Aberdeen Journal*, Jan. 24, 1844, 1; "Book Review," *Monthly Review*, Sept. 1843, 82; "A Narrative of the Discoveries on the North Coast of America; Effected by the Officers of the Hudson's Bay Company during the Years 1836–1839," *Examiner*, Aug. 26, 1843, 532.

52. For more on King's life and conflicts surrounding the HBC, see Hugh N. Wallace, *The Navy, the Company, and Richard King: British Exploration in the Canadian Arctic, 1829–1860* (Montreal: McGill-Queen's University Press, 1980); Efram Sera-Shriar, "Arctic Observers: Richard King,

Monogenism and the Historicisation of Inuit through Travel Narratives," *Studies in History and Philosophy of Science, Part C: Studies in History and Philosophy of Biological and Biomedical Sciences* 51 (June 2015): 23–31.

53. Simpson, *Narrative of the Discoveries*, 221.

54. Hilary E. Wyss, *Writing Indians: Literacy, Christianity, and Native Community in Early America*, pbk. ed. (Amherst: University of Massachusetts Press, [2000] 2003), 126, 143; Gregory Eiselein, *Literature and Humanitarian Reform in the Civil War Era* (Bloomington: Indiana University Press, 1996), 141–44.

55. Simpson, *Narrative of the Discoveries*, 73–74, 228, 73.

56. [Egerton Francis], "ART. V.—Narrative of the Discoveries on the North Coast of America, Effected by the Officers of the Hudson's Bay Company, during the Years 1836–39," *Quarterly Review*, Dec. 1843, 117. This article was published anonymously by Egerton Francis (1800–1857) and was included in his *Essays on History, Biography, Geography, Engineering &c. Contributed to the Quarterly Review by the Late Earl of Ellesmere* (London: John Murray, 1858).

57. "Discoveries on the North Coast of America," *Chambers's Edinburgh Journal*, Sept. 1844, 277.

58. Gregory P. Marchildon, *The Early Northwest* (Regina, SK: University of Regina Press, 2008), 181.

59. Simpson, *Narrative of the Discoveries*, 146–47.

60. Simpson, *Narrative of the Discoveries*, 149.

61. Simpson, *Narrative of the Discoveries*, xi, xii.

62. "Thomas Simpson, Esq.," *Gentleman's Magazine and Historical Review, July 1856–May 1868*, Nov. 1840, 548; "Our Weekly Gossip," *Athenaeum*, Sept. 5, 1840, 701.

63. Alexander Simpson, preface to Simpson, *Narrative of the Discoveries on the North Coast of America*, xvii (quote); Simpson, *Life and Travels of Thomas Simpson*.

64. "A. Narrative of the Discoveries on the North Coast of America; Effected by the Officers of the Hudson's Bay Company during the Years 1836–1839," 532; "Narrative of the Discoveries on the North Coast of America, Effected by the Officers of the Hudson's Bay Company, during the Years 1836–9," *Critic of Literature, Art, Science, and the Drama, 1843–1844*, Feb. 1844, 85.

65. S.R, "ART. VI.—The Journal of the Royal Geographical Society of London.," *London and Westminster Review, Apr. 1836–Mar. 1840* 31, no. 2 (Aug. 1838): 389.

66. Simpson, *Narrative of the Discoveries*, 156, 190.

67. Harold Adams Innis, "The Importance of Staple Products" (1930), in William Thomas Easterbrook and Mel Watkins, eds., *Approaches to Canadian Economic History: A Selection of Essays* (Toronto: Carleton University Press, [1984] 2003), 16–19. See also Innis, *The Fur Trade in Canada: An Introduction to Canadian Economic History*, rev. ed. (Toronto: University of Toronto Press, 1999); Robin Neill, *A History of Canadian Economic Thought*, Routledge History of Economic Thought Series (London: Routledge, 1991).

68. Simpson, *Narrative of the Discoveries*, 12–13.

69. See Perry Adele, "Designing Dispossession: The Select Committee on the Hudson's Bay Company, Fur-Trade Governance, Indigenous Peoples and Settler Possibility," in *Indigenous Communities and Settler Colonialism: Land Holding, Loss and Survival in an Interconnected World*, ed. Zoë Laidlaw and Alan Lester (New York: Palgrave Macmillan, 2015), 158.

70. Simpson, *Narrative of the Discoveries*, 132.

Chapter 3: The Lost Franklin Expedition and New Opportunities for Arctic Exploration
Epigraph: Richard King, "The Arctic Expeditions," *Athenaeum*, Dec. 11, 1847, 1273.

1. The total number of expeditions varies depending on the historical source. For a survey of the

expeditions, see W. Gillies Ross, "The Type and Number of Expeditions in the Franklin Search 1847–1859," *Arctic* 55, no. 1 (2002): 57–69.

2. Jane Franklin to Sir James Graham, Jan. 20, 1854, MS 1100/7;D (quote); Jane Franklin to Admiralty, Jan. 26, 1854, MS 248/212/6, both in SPRI.

3. "Multiple News Items," *Morning Post*, Oct. 26, 1849.

4. Robert Murchison, quoted in Francis Leopold, "Discoveries by the Late Expedition in Search of Sir John Franklin and His Party," *Proceedings of the Royal Geographical Society of London* 30 (1860): 13.

5. John Barrow, "Proposal for an Attempt to Complete the Discovery of a North-West Passage," Dec. 1844, reference no. GB 117 MM/10/172, Royal Society Archives, London.

6. John Barrow, quoted in Charles Richard Weld, *Arctic Expeditions* (London: John Murray, 1850), 18.

7. Barrow quoted in Weld, *Arctic Expeditions*, 20–22.

8. There has been a continued and significant amount of both scholarly and nonacademic attention devoted to the Franklin expedition and its rescue missions. See for example: Anthony Brandt, *The Man Who Ate His Boots: Sir John Franklin and the Tragic History of the Northwest Passage* (New York: Random House, 2011); David C. Woodman, *Unravelling the Franklin Mystery: Inuit Testimony*, 2nd ed. (Montreal: McGill-Queen's Press, 2015); Laurie Garrison, "Virtual Reality and Subjective Responses: Narrating the Search for the Franklin Expedition through Robert Burford's Panorama," *Early Popular Visual Culture* 10, no. 1 (2012): 7–22; David Murphy, *The Arctic Fox: Francis Leopold-McClintock* (Toronto: Dundurn, 2004); Martin W. Sandler, *Resolute: The Epic Search for the Northwest Passage and John Franklin, and the Discovery of the Queen's Ghost Ship* (New York: Sterling Publishing, 2008); C. Stuart Houston and John Richardson, *Arctic Ordeal: The Journal of John Richardson, Surgeon-Naturalist with Franklin, 1820–1822* (Montreal: McGill-Queen's Press, 1994); Russell A. Potter, *Finding Franklin: The Untold Story of a 165-Year Search* (Montreal: McGill-Queen's Press, 2016); Ross, "Type and Number of Expeditions"; John Geiger and Owen Beattie, *Frozen in Time* (London: Bloomsbury, [1987] 2012); Jeffrey Blair Latta, *The Franklin Conspiracy: An Astonishing Solution to the Lost Arctic Expedition* (Toronto: Dundurn Press, 2001); Scott Cookman, *Ice Blink: The Tragic Fate of Sir John Franklin's Lost Polar Expedition* (New York: John Wiley & Sons, 2001); John Geiger and Alanna Mitchell, *Franklin's Lost Ship: The Historic Discovery of HMS Erebus* (Toronto: HarperCollins, 2015); W. Gillies Ross, "The Admiralty and the Franklin Search," *Polar Record* 40, no. 4 (Oct. 2004): 289–301; Liz Cruwys, "Henry Grinnell and the American Franklin Searches," *Polar Record* 26, no. 158 (July 1990): 211–16; Shane McCorristine and Victoria Herrmann, "The 'Old Arctics': Notices of Franklin Search Expedition Veterans in the British Press, 1876–1934," *Polar Record* 52, no. 2 (Mar. 2016): 215–29; W. Gillies Ross, "False Leads in the Franklin Search," *Polar Record* 39, no. 2 (Apr. 2003): 131–60; William Barr, "Searching for Franklin Where He Was Ordered to Go: Captain Erasmus Ommanney's Sledging Campaign to Cape Walker and beyond, Spring 1851," *Polar Record* 52, no. 4 (July 2016): 474–98; Janice Cavell, "Going Native in the North: Reconsidering British Attitudes during the Franklin Search, 1848–1859," *Polar Record* 45, no. 1 (Jan. 2009): 25–35.

9. Ted Binnema, *Enlightened Zeal: The Hudson's Bay Company and Scientific Networks, 1670–1870* (Buffalo, NY: University of Toronto Press, 2014), 160.

10. Richardson may have suffered a heart attack during this expedition, which caused him to return early. See, for example, D. A. Stewart, "Sir John Richardson Surgeon, Physician, Sailor, Explorer, Naturalist, Scholar," *British Medical Journal* 1, no. 3654 (1931): 110–12. However, Jane Franklin noted in a letter to Rae, that Richardson looked healthier upon his return than before departing for their expedition. Much of the surviving correspondence is held in the John Rae Collection, SPRI. Jane Franklin to John Rae (copy of the letter), May 3, 1850, MS 248/198;D, SPRI.

11. John Richardson, *Arctic Searching Expedition: A Journal of a Boat-Voyage through Rupert's Land and the Arctic Sea, in Search of the Discovery Ships under Command of Sir John Franklin. With an Appendix on the Physical Geography of North America*, 2 vols. (London: Longman, Brown, Green and Longmans, 1851).

12. Rae outlined the provisions used on his expeditions in his report: John Rae, *Details of Provisions Used on Arctic Expeditions*, 1846, MS 787/9/1–2;D, SPRI.

13. Robert Jameson, "Literary and Scientific Intelligence," *Edinburgh Magazine and Literary Miscellany*, 1817, 367–69.

14. Adrian Desmond and James Moore, *Darwin's Sacred Cause: Race, Slavery and the Quest for Human Origins* (London: Penguin Books, 2009), 28. See also R. E. Johnson, "Biography—Richardson, Sir John," *Dictionary of Canadian Biography*, vol. 9, *1861–1870*, http://www.biographi.ca/en/bio.php?id_nbr=4670 (accessed July 22, 2016).

15. The main biography of Rae is Kenneth McGoogan, *Fatal Passage: The Story of John Rae, the Artic Hero Time Forgot* (New York: Carroll & Graf, 2002).

16. John Rae, *Narrative of an Expedition to the Shores of the Arctic Sea, in 1846 and 1847* (London: T. & W. Boone, 1850), 21.

17. Jane Franklin to John Rae.

18. Richardson, *Fauna Boreali-Americana*; Hooker, *Flora Boreali-Americana*.

19. Richardson, *Arctic Searching Expedition*. Trevor Levere has pointed out that this was linked to the nascent discipline of biogeography in his *Science and the Canadian Arctic: A Century of Exploration, 1818–1918* (Cambridge: Cambridge University Press, 2004), 176–77. For more on biogeography and collecting, see Janet Browne, *The Secular Ark: Studies in the History of Biogeography* (New Haven, CT: Yale University Press, 1983). For my understanding of Humboldtian science, I am particularly informed by Suzanne Zeller, "Humboldt and the Habitability of Canada's Great Northwest," *Geographical Review* 96, no. 3 (2006): 387.

20. Werner developed a theory of the stratification of the Earth's crust that had as its basis the theory that the Earth had been fully covered by an ocean that then gradually receded, termed Neptunism. Werner's scheme created an immediate controversy about the origin of basalt that became the foundation of the Neptunist-Plutonist controversy in eighteenth- and early nineteenth-century geological circles. Trevor Levere has pointed out that Richardson's belief in Neptunism was why he wrongly described the basalt in the Copper Mountains as aqueous in origin; Levere, *Science and the Canadian Arctic*, 109.

21. Rachel Laudan, *From Mineralogy to Geology: The Foundations of a Science, 1650–1830* (Chicago: University of Chicago Press, 1987), 104–5; Mott T. Greene, *Geology in the Nineteenth Century: Changing Views of a Changing World* (Ithaca, NY: Cornell University Press, 1982), 43–44.

22. Richardson, *Arctic Searching Expedition*, 2:162.

23. Richardson, *Arctic Searching Expedition*, 1:116, 152, 2:167. The main resource for the significance of the Devonian system is Martin J. S. Rudwick, *The Great Devonian Controversy: The Shaping of Scientific Knowledge among Gentlemanly Specialists* (Chicago: University of Chicago Press, 1988). Richardson, *Arctic Searching Expedition*, 1:327.

24. Susan Sheets-Pyenson, "'Pearls before Swine': Sir William Dawson's Bakerian Lecture of 1870," *Notes and Records of the Royal Society of London* 45, no. 2 (1991): 182.

25. Richardson, *Arctic Searching Expedition*, 2:167n14.

26. Alexander von Humboldt, *Des lignes isothermes et de la distribution de la chaleur sur le globe* (Paris: Perronneau, 1817).

27. Richardson, *Arctic Searching Expedition*, 2:258; British Association for the Advancement of Science, *Report of the 17th Meeting of the British Association for the Advancement of Science (Oxford)* (London: Taylor & Francis, 1848), 373–76.

28. Michael Dettelbach, "The Face of Nature: Precise Measurement, Mapping, and Sensibility in the Work of Alexander von Humboldt," *Studies in History and Philosophy of Science Part C: Studies in History and Philosophy of Biological and Biomedical Sciences* 30, no. 4 (1999): 486; Zeller, "Humboldt and the Habitability of Canada's Great Northwest," 392. See also Katharine Anderson, *Predicting the Weather: Victorians and the Science of Meteorology* (Chicago: University of Chicago Press, 2005), 257; and Paul N. Edwards, *A Vast Machine: Computer Models, Climate Data, and the Politics of Global Warming* (Cambridge, MA: MIT Press, 2010), 31.

29. Richardson, *Arctic Searching Expedition*, 2:271.

30. Richardson, *Arctic Searching Expedition*, 2:275.

31. Richardson, *Arctic Searching Expedition*, 2:55–59.

32. Richardson, *Arctic Searching Expedition*, 2:18, 1:241. It appears that Richardson used *Tinnè/Chepewyan* as an umbrella term for several groups of peoples in Western Canada. The Chipewyan are an aboriginal Dene people.

33. "Narrative of an Expedition to the Shores of the Arctic Sea in 1846 and 1847," *Athenaeum*, July 1850, 784.

34. Richard King, *The Franklin Expedition from First to Last* (London: John Churchill, 1855), 124.

35. Gillian Beer, *Open Fields: Science in Cultural Encounter* (Oxford: Oxford University Press, 1999), 46; Levere, *Science and the Canadian Arctic*, 202.

36. See, for example, "The Fate of Sir John Franklin," *Illustrated London News*, Oct. 28, 1854; "The Fate of Franklin," *Morning Post*, Oct. 23, 1854; "Probable Fate of Sir John Franklin's Party," *Morning Chronicle*, Oct. 23, 1854; "The Fate of Sir John Franklin," *Daily News*, Oct. 23, 1854; "The Arctic Expedition," *Times*, Oc. 23, 1854; "Multiple News Items," *Standard*, Oct. 23, 1854.

37. "Fate of Sir John Franklin," *Daily News*.

38. Charles Dickens, "The Lost Arctic Voyagers," part 1, *Household Words*, Dec. 2, 1854, 361–65; Charles Dickens, "The Lost Arctic Voyagers," part 2, *Household Words*, Dec. 9, 1854, 385–93. See also Priti Joshi, "Race," in *Charles Dickens in Context*, ed. Sally Ledger and Holly Furneaux (Cambridge: Cambridge University Press, 2011), 292–300; Alana Lentin, *Racism and Ethnic Discrimination* (New York: The Rosen Publishing Group, 2011), 55–56; Sadiah Qureshi, *Peoples on Parade: Exhibitions, Empire, and Anthropology in Nineteenth-Century Britain* (Chicago: University of Chicago Press, 2011), 177–81; Grace Moore, *Dickens and Empire: Discourses of Class, Race and Colonialism in the Works of Charles Dickens* (Aldershot, UK: Ashgate, 2004).

39. Dickens, "Lost Arctic Voyagers," part 1, 362.

40. See, for example, Efram Sera-Shriar, *The Making of British Anthropology, 1813–1871* (London: Pickering and Chatto; Pittsburgh: University of Pittsburgh Press, 2013); Efram Sera-Shriar, "Arctic Observers: Richard King, Monogenism and the Historicisation of Inuit through Travel Narratives," *Studies in History and Philosophy of Science Part C: Studies in History and Philosophy of Biological and Biomedical Sciences* 51 (June 2015): 23–31; George W. Stocking, *The Ethnographer's Magic and Other Essays in the History of Anthropology* (Madison: University of Wisconsin Press, 1992); Stocking, *Observers Observed: Essays on Ethnographic Fieldwork* (Madison: University of Wisconsin Press, 1984); Henrika Kuklick, "Personal Equations: Reflections on the History of Fieldwork, with Special Reference to Sociocultural Anthropology," *Isis* 102, no. 1 (2011): 1–33; Richard C. Powell, "Becoming a Geographical Scientist: Oral Histories of Arctic Fieldwork," *Transactions of the Institute of British Geographers*, n.s., 33, no. 4 (Oct. 2008): 548–65.

41. Daniel Panneton and Leslie H. Neatby, "John Rae," *The Canadian Encyclopedia*, http://www.thecanadianencyclopedia.com/en/article/john-rae/ (accessed Dec. 19, 2016).

42. Jane Franklin to John Rae.

43. Jane Franklin to the Admiralty, Nov. 29, 1854, MS 248/212/8, SPRI.

44. Jane Franklin to Turner (solicitor), May 31, 1855, MS 248/178;D, SPRI.

45. Janice Cavell, "Publishing Sir John Franklin's Fate: Cannibalism, Journalism, and the 1881 Edition of Leopold McClintock's The Voyage of the 'Fox' in the Arctic Seas," *Book History* 16, no. 1 (Oct. 2013): 178.

46. King, *Franklin Expedition from First to Last*, 133.

47. Hugh N. Wallace, *The Navy, the Company, and Richard King: British Exploration in the Canadian Arctic, 1829–1860* (Montreal: McGill-Queen's University Press, 1980), 148.

48. King, *Franklin Expedition from First to Last*, 131.

49. Elisha Kent Kane, *Arctic Explorations: The Second Grinnell Expedition in Search of Sir John Franklin, 1853, '54, '55*, 2 vols. (Philadelphia: Childs & Peterson, 1857).

50. Henry Grinnell to Jane Franklin, Feb. 10, 1852, MS 248/414/33, SPRI.

51. Many of these letters are held at SPRI, folder reference GB 15 Henry Grinnell. See, for example, Henry Grinnell to Jane Franklin, Mar. 4, 1850 to Nov. 22, 1873, MS 248/414/2–11;D, SPRI.

52. Kane, *Arctic Explorations*, 2:301.

53. Potter, *Finding Franklin*, 84; McGoogan, *Fatal Passage*, 230; Janice Cavell, *Tracing the Connected Narrative: Arctic Exploration in British Print Culture, 1818–1860* (Toronto: University of Toronto Press, 2008), 216.

54. No known vital dates.

55. Rae, *Narrative of an Expedition*, 126.

56. No known vital dates.

57. Rae, *Narrative of an Expedition*, 88.

58. Potter, *Finding Franklin*, 87.

59. "Sir John Richardson's Arctic Searching Expedition," *Spectator* Nov. 15, 1851, 1096; "Narrative of an Expedition to the Shores of the Arctic Sea in 1846 and 1847," *Athenaeum*; "Narrative of an Expedition to the Shores of the Arctic Sea, in 1846 and 1847," *Quarterly Review*, Mar. 1853, 386–421; "Arctic Searching Expedition," *Examiner*, Dec. 6, 1851, 772.

60. "Sir John Richardson's Arctic Expedition," *Dublin University Magazine*, Apr. 1852, 458–76; "Arctic Searching Expedition," *Examiner*; "Arctic Searching Expedition: A Journal of a Boat Voyage through Rupert's Land and the Arctic Sea, in Search of the Discovery Ships under Command of Sir John Franklin. With an Appendix on the Physical Geography of North America.," *North British Review* 16, no. 32 (Feb. 1852): 445–89; "Arctic Searching Expedition: A Journal of a Boat-Voyage through Rupert's Land and the Arctic Sea, in Search of the Discovery Ships under Command of Sir John Franklin. With an Appendix on the Physical Geography of North America," *Athenaeum*, Nov. 29, 1851, 1246–47; "Narrative of an Expedition to the Shores of the Arctic Sea, in 1846 and 1847," *Quarterly Review*, 397.

61. Jane Franklin, Letter Book, Nov. 23, 1856, to Jan. 10, 1857, July 1856, MS 248/114;BJ, 74; John Rae to John Richardson, July 30, 1850, MS 909/6/1; John Rae to John Richardson," Sept. 20, 1849, MS 909/4/1, all in SPRI.

62. John Rae to John Richardson, Aug. 18, 1855, MS 1503/50/32, Richardson-Voss Collection, SPRI; Wallace, *Navy, the Company, and Richard King*, 148; McGoogan, *Fatal Passage*, 234; Sophia Cracroft to John Richardson, Mar. 1, 1856, MS 1503/52/39;RV, Richardson-Voss Collection, SPRI; Franklin, Letter Book, 29.

63. The letter, Roderick Murchison et al., "Memorial to the Right Hon. Viscount Palmerston, M.P., G.C.B," June 5, 1856, was included in the appendix of Francis Leopold M'Clintock, *The Voyage of the "Fox" in the Arctic Seas: A Narrative of the Discovery of the Fate of Sir John Franklin and His Companions* (London: John Murray, 1859), 361–65.

64. M'Clintock, *Voyage of the "Fox,"* 361; Robert Inglis, quoted in "Imperial Parliament," *Standard*, June 13, 1849, 3; M'Clintock, *Voyage of the "Fox,"* 11.

65. Marc Rothenberg, "Making Science Global? Coordinated Enterprises in Nineteenth-Century Science," in *Globalizing Polar Science: Reconsidering the International Polar and Geophysical Years*, ed. Roger D. Launius, James Rodger Fleming, and David H. DeVorkin, Palgrave Studies in the History of Science and Technology (New York: Palgrave Macmillan, 2010), 27. For more on the question of international collaboration and the magnetic crusade, see Jessica Ratcliff, *The Transit of Venus Enterprise in Victorian Britain* (London: Pickering and Chatto; Pittsburgh: University of Pittsburgh Press, 2008), 24–25; Maurice Crosland, *Science under Control: The French Academy of Sciences 1795–1914* (New York: Cambridge University Press, 1992), 377; John Cawood, "Terrestrial Magnetism and the Development of International Collaboration in the Early Nineteenth Century," *Annals of Science* 34, no. 6 (Nov. 1977): 551–87; Rothenberg, "Making Science Global?"; Christopher Carter, "Going Global in Polar Exploration: Nineteenth-Century American and British Nationalism and Peacetime Science," in Launius, Fleming, and DeVorkin, *Globalizing Polar Science*, 86–105.

66. Francis Leopold McClintock to Sir James Clark Ross, Feb. 13, 1855, MS 1226/17/2; Francis Leopold McClintock to Sir James Clark Ross, Mar. 15, 1855, MS 1226/17/3; Francis Leopold McClintock to Jane Franklin, Apr. 18, 1857, MS 248/439/23, all in SPRI.

67. M'Clintock, *Voyage of the "Fox"*; Carl Petersen, *Den Sidste Franklin-Expedition med "Fox," Capt. M'Clintock, ved Carl Petersen* (Copenhagen: Fr. Woldikes Forlagsboghandel, 1860).

68. The main biography of Carl Petersen is Nils Aage Jensen, *Carl—Polarfarer* (Copenhagen: Lindhardt og Ringhof, 2014).

69. Petersen, *Den Sidste Franklin-Expedition med "Fox,"* 84; Francis Leopold McClintock to Richard Collinson," July 3, 1858, MS 248/439/7, SPRI.

70. Francis Leopold McClintock to Jane Franklin, Oct. 19, 1864, MS 248/439/35, SPRI; Jensen, *Carl—Polarfarer*, 12–13. See also "Indlandet," *Vestslesvigsk Tidende*, Feb. 15, 1860, 2; "Den Sidste Franklin-Expedition med 'Fox,'" *Lolland-Falsters Stifts-Tidende*, May 26, 1860, 2; "Literatur," *Fyens Stiftstidende*, May 10, 1860, 1.

71. No vital dates.

72. Petersen, *Den Sidste Franklin-Expedition med "Fox,"* 14–15.

73. Kane, *Arctic Explorations*, 2:318; Francis Leopold McClintock to Jane Franklin, July 21, 1857, MS 248/439/27, SPRI; Francis Leopold McClintock to Jane Franklin, June 3, 1858.

74. Svend Thorsen and Tage Kaarsted, *De Danske Ministerier: Et Hundred Politisk-Historiske Biografier [Udg. af Pensionsforsikringsanstalten i Anledning af dens 50 Års Jubilaeum]* (Nyt Nordisk Forlag, 1967), 119–20; Jensen, *Carl—Polarfarer*, 341–42. See also Roderick Impey Murchison, *Address to the Royal Geographical Society of London; Delivered at the Anniversary Meeting, May 25th, 1857* (London: W. Clowes and Sons, 1857), 113n.

75. No known vital dates.

76. McClintock to Franklin, June 3, 1858; M'Clintock, *Voyage of the "Fox,"* 25–26.

77. John Rae to Unknown, draft of a letter, Dec. 15, 1877, MS 787/4;D, SPRI.

78. Petersen, *Den Sidste Franklin-Expedition med "Fox,"* 22; "Postnyheder," *Kalundborg Avis*, Oct. 1, 1859, 1; Petersen, *Den Sidste Franklin-Expedition med "Fox,"* 8. See also Cavell, "Publishing Sir John Franklin's Fate," 160.

79. Petersen, *Den Sidste Franklin-Expedition med "Fox,"* 21.

80. Sophia Cracroft to Henry Grinnell, copy of letter, Nov. 21, 1856, MS 248/250/7, SPRI.

81. Sophia Cracroft to Henry Grinnell, draft or copy of letter, Apr. 28, 1857, MS 248/250/12, SPRI.

82. Sophia Cracroft to Henry Grinnell, draft or copy of letter, Apr. 10, 1857, MS 248/250/11, SPRI; Cracroft to Grinnell, Apr. 28, 1857.

83. "Udenlandske Efterretninger," *Den til Forsendelse med de Kongelige Brevposter Privilegerede Berlingske Politiske og Avertissementstidende,* Oct. 28, 1854, 2; "Repulse-Bay," *Dannevirke,* Oct. 31, 1854, 2; "Udlandet," *Fyens Stiftstidende,* Oct. 30, 1854, 1; "Expeditionerne efter Sir John Franklin," *Sjællands-Posten,* Oct. 31, 1854, 2.

84. Petersen, *Den Sidste Franklin-Expedition med "Fox,"* 232–33; M'Clintock, *Voyage of the "Fox,"* 276.

85. See for example, Woodman, *Unravelling the Franklin Mystery,* 53.

86. Francis Leopold McClintock to Jane Franklin, Aug. 3, 1858, MS 248/439/33, SPRI.

87. William Barr, "The Use of Dog Sledges during the British Search for the Missing Franklin Expedition in the North American Arctic Islands, 1848–59," *Arctic* 62, no. 3 (2009): 261.

88. H. B., "Den Sidste Franklin-Expedition med 'Fox,'" *Fædrelandet,* Dec. 22, 1860, 1.

89. Petersen, *Den Sidste Franklin-Expedition med "Fox,"* 7.

90. Petersen, *Den Sidste Franklin-Expedition med "Fox,"* 233.

91. Russell A. Potter, "Introduction: Exploration and Sacrifice: The Cultural Logic of Arctic Discovery," in *Arctic Exploration in the Nineteenth Century: Discovering the Northwest Passage,* ed. Frédéric Regard (London: Pickering and Chatto; Pittsburgh: University of Pittsburgh Press, 2015), 6.

92. "The Arctic Expedition," *North Devon Journal,* May 9, 1850.

Chapter 4: From Science in the Arctic to Arctic Science

Epigraph: Clements Robert Markham, *The Threshold of the Unknown Region* (London: Sampson Low, Marston, Low, and Searle, 1873), 335; [Clements Robert Markham], "The Arctic Campaign of 1873," *Ocean Highways: The Geographical Record* 1, no. 3 (1874): 91.

1. See, for example, Markham, *Threshold of the Unknown Region,* 335. The issue of Arctic exploration was repeatedly treated in the *Geographical Magazine,* edited by Marham, under the section "Arctic Region." For example, the 1875–1876 expedition was discussed in several instances in 1876.

2. Marc Rothenberg, "Making Science Global? Coordinated Enterprises in Nineteenth-Century Science," in *Globalizing Polar Science: Reconsidering the International Polar and Geophysical Years,* ed. Roger D. Launius, James Rodger Fleming, and David H. DeVorkin, Palgrave Studies in the History of Science and Technology (New York: Palgrave Macmillan, 2010), 23; Jürgen Osterhammel and Niels P. Petersson, *Globalization: A Short History,* trans. Dona Geyer (Princeton, NJ: Princeton University Press, 2005), 26; C. A. Bayly et al., "On Transnational History," AHR Conversation, *American Historical Review* 111, no. 5 (Dec. 1, 2006): 1446; Sujit Sivasundaram, "Sciences and the Global: On Methods, Questions, and Theory," *Isis* 100, no. 1 (Mar. 2010): 146, 154.

3. Clements Robert Markham to Robert Brown of Campster, London, Dec. 7, 1880, MS 441/9/1– 42 D, SPRI.

4. No known vital dates. He is believed to have passed away soon after the Second International Polar Year between 1932 and 1933.

5. "Indland," *Aarhuus Stifts-Tidende,* Oct. 26, 1877.

6. I am indebted to conversation with John Woitkowitz, who is undertaking research on theories on the Open Polar Sea. Woitkowitz, "Science, Networks, and Knowledge Communities: August

Petermann and the Construction of the Open Polar Sea," paper presented at the annual meeting of the Canadian Historical Association, Vancouver, June 5, 2019; Woitkowitz and Nanna Kaalund, "August Petermann, Elisha Kent Kane, and the Making of the Open Polar Sea," paper presented at the Science Museum Seminar Series, London, June 18, 2019.

7. Hans Hendrik, *Memoirs of Hans Hendrik : The Arctic Traveller, Serving under Kane, Hayes, Hall and Nares, 1853–1876*, ed. George Stephens, trans. Hinrich Rink (London: Trübner, 1878), 3; Preben Andersen, "Herrnhutterne i Grønland," *Tidsskriftet Grønland*, no. 2 (1969): 50–64; Mads Lidegaard, "Hans Hendrik fra Fiskenæsset," *Grønland* 8 (Aug. 1968): 249–50. The main biography of Suersaq is Jan Løve, *Hans Hendrik og Hans Ø: Beretningen om Hans Hendrik og de to Hans Øer* (Copenhagen: Det Grønlandske Selskab, 2016).

8. For an account of the Kane expedition with a focus on the scientific aspects, see David Chapin, *Exploring Other Worlds: Margaret Fox, Elisha Kent Kane, and the Antebellum Culture of Curiosity* (Amherst: University of Massachusetts Press, 2004), 54–74. For a detailed account of Hall's expedition, see Bruce B. Henderson, *Fatal North: Adventure and Survival aboard USS Polaris, the First U.S. Expedition to the North Pole* (New York: New American Library, 2001).

9. Henry Rawlinson, quoted in "Arctic Exploration," *Times*, Dec. 17, 1872, 8. See also Philip N. Cronenwett, "British Arctic Expedition, 1875–1876," in *Encyclopedia of the Arctic*, ed. Mark Nuttall (New York: Routledge, 2012), 277.

10. [John Thadeus Delane?], "The Reasons Which Make It Desirable to Despatch," *Times*, Dec. 18, 1872, 9; An Arctic Officer, "Polar Exploration," letter to the editor, *Times*, Dec. 26, 1872, 8.

11. John Rae, "Arctic Exploration," letter to the editor, *Times*, Dec. 28, 1872, 3; Arctic Officer, "Polar Exploration," 8.

12. Rae, "Arctic Exploration," 3; See also Rae's follow-up letter: John Rae and John C. Wells, "Arctic Exploration," letter to the editor, *Times*, Dec. 31, 1872, 10; John Rae, lecture fragment, n.d., MS 787/12/1;D, SPRI.

13. John Campion Wells and B. Leigh-Smith, "Arctic Exploration," *Times*, Nov. 19, 1872; Rae and Wells, "Arctic Exploration"; John Campion Wells, *The Gateway to the Polynia: A Voyage to Spitzbergen* (London: Henry S. King, 1873).

14. Royal Geographical Society of Great Britain, ed., "Sessions 1872–73," *Proceedings of the Royal Geographical Society of London* (1873): 77. See also, Michael S. Reidy, Gary R. Kroll, and Erik M. Conway, *Exploration and Science: Social Impact and Interaction* (Santa Barbara, CA: ABC-CLIO, 2007), 96–97.

15. Marvin Swartz, *Politics of British Foreign Policy in the Era of Disraeli and Gladstone* (New York: St. Martin's Press, 1985), 12.

16. Great Britain. Royal Navy. Admiralty, *Arctic Expedition: Papers and Correspondence Relating to the Equipment and Fitting Out of the Arctic Expedition of 1875, Including Report of the Admiralty Arctic Committee; Presented to Both Houses of Parliament by Command of Her Majesty* (London: George Edward Eyre and William Spottiswoode, 1875), 17; Hendrik, *Memoirs of Hans Hendrik*, 84.

17. Elizabeth Bella, "British Arctic Expedition," in *Antarctica and the Arctic Circle: A Geographic Encyclopedia of the Earth's Polar Regions*, ed. Andrew Jon Hund (Santa Barbara, CA: ABC-CLIO, 2014), 162.

18. George Strong Nares, *Narrative of a Voyage to the Polar Sea: During 1875–6 in H. M. Ships 'Alert' and 'Discovery,'* 2 vols. (London: Slow, Marston, Searle, & Rivington, 1878); Nares, *The Official Report of the Recent Arctic Expedition* (London: John Murray, 1876), 72; Hendrik, *Memoirs of Hans Hendrik*, 89.

19. Nares, *Official Report of the Recent Arctic Expedition*, 70; Nares, *Narrative of a Voyage*, 2:82, 96–97, 111.

20. Hendrik, *Memoirs of Hans Hendrik*, 97.

21. Hendrik, *Memoirs of Hans Hendrik*, 86, 91, 89.

22. Julie Cruikshank, *Do Glaciers Listen? Local Knowledge, Colonial Encounters, and Social Imagination* (Vancouver: UBC Press, 2010), 76–77; Sophus Theodor Krarup-Smith to D. L. Braine, Sept. 10, 1873, Area File 7, National Archives, Washington, DC.

23. Frantz Fanon, *The Wretched of the Earth*, new trans. by Richard Philcox (New York: Grove Press, [1961] 2007), For polemic takes on Fanon's theories, see Munyaradzi Hwami, "Frantz Fanon and the Problematic of Decolonization: Perspectives on Zimbabwe," *African Identities* 14, no. 1 (Jan. 2, 2016): 19–37, Michael Groden, Martin Kreiswirth, and Imre Szeman, eds., *Contemporary Literary and Cultural Theory: The Johns Hopkins Guide* (Baltimore: Johns Hopkins University Press, 2012), 155–57; Michael D. Wilson, *Writing Home: Indigenous Narratives of Resistance* (East Lansing: Michigan State University Press, 2008), x; Gary Wilder, "Race, Reason, Impasse: Cesaire, Fanon, and the Legacy of Emancipation," *Radical History Review* 90, no. 1 (Aug. 2004): 31–61. For an overview of postcolonial theories, see Elleke Boehmer and Bart Moore-Gilbert, "Introduction to Special Issue: Postcolonial Studies and Transnational Resistance," *Interventions* 4, no. 1 (Jan. 2002): 7–21. For further reflections on translation, see also Cruikshank, *Do Glaciers Listen?*, 66.

24. Hendrik, *Memoirs of Hans Hendrik*, 89–91.

25. Hinrich Rink, Introduction to Hendrik, *Memoirs of Hans Hendrik*. I have not been able to locate these notes, and it is uncertain if they still exist.

26. Rink, Introduction, 20; Isidore Diala, "Colonial Mimicry and Postcolonial Re-Membering in Isidore Okpewho's Call Me by My Rightful Name," *Journal of Modern Literature* 36, no. 4 (2013): 77–95.

27. The most detailed biographies of Hinrich and Signe Rink are Knud Oldendow, *Grønlændervennen Hinrich Rink: Videnskabsmand, Skribent Og Grønlandsadministrator*, Det Grønlandske Selskabs Skrifter 18 (Copenhagen: Det Grønlandske Selskab, 1955); Ole Marquardt, "Between Science and Politics: The Eskimology of Hinrich Johannes Rink," in *Early Inuit Studies: Themes and Transitions, 1850s–1980s*, ed. Igor Krupnik (Washington, DC: Smithsonian Institution Press, 2016), 35–54. See also George Nellemann, "Hinrich Rink and Applied Anthropology in Greenland in the 1860's," *Human Organization* 28, no. 2 (June 1969): 166–74; K. J. V. Steenstrup, "Dr. Phil. Hinrich Johannes Rink," *Geografisk Tidsskrift* 12 (Jan. 1894): 162–66; Bodil Kaalund, *The Art of Greenland: Sculpture, Crafts, Painting*, trans. Kenneth Tindall (Berkeley: University of California Press, 1983), 164; Pamela R. Stern, *Daily Life of the Inuit* (Santa Barbara CA: Greenwood, 2010), 129; and Axel Garboe, *Geologiens Historie i Danmark: Forskere og Resultater* (Copenhagen: C. A. Reitzel, 1961), 2:216.

28. Steenstrup, "Dr. Phil. Hinrich Johannes Rink"; Hinrich Rink, *Om Grønlænderne, deres Fremtid, og de til deres Bedste Sigtende Foranstaltninger* (Copenhagen: Høst, 1882); Hugo Hørring, *Bemærkninger til Justitsraad, Dr. Phil. H. Rinks Skrift: Om Grønlænderne M.m.* (Copenhagen: Kommission hos GECGad, 1882); Hugo Hørring, *Fortsatte Bemærkninger til Justitsraad Dr. Phils. H. Rinks Skrift: Om Grønlænderne M.m.* (Copenhagen, 1883); Rink, "The Recent Danish Explorations in Greenland and Their Significance as to Arctic Science in General," *Proceedings of the American Philosophical Society* 22, no. 120 (1885): 280–96.

29. Hinrich Rink, "Nogle Bemærkninger om de Nuværende Grønlænderes Tilstand," *Geografisk Tidsskrift* 1 (Jan. 1877): 29; Rink, Introduction, 5.

30. Hendrik, *Memoirs of Hans Hendrik*, 89–91; "Book Review," *Athenaeum*, Oct. 26, 1878, 527.

31. The concept of implicit ethnography and implicit understanding is described in detail in the introduction to Stuart B. Schwartz, ed., *Implicit Understandings: Observing, Reporting and Reflecting on the Encounters between Europeans and Other Peoples in the Early Modern Era* (Cambridge: Cambridge University Press, 1994), 1–20; Hendrik, *Memoirs of Hans Hendrik*, 89–90.

32. Rink, Introduction, 5.

33. "Book Review," *Athenaeum*, 527; Sophus Theodor Krarup-Smith to D. L. Braine, Sept. 10, 1879, 300258, National Archives, Washington, DC.

34. Hinrich Rink, Introduction, 2.

35. George Stephens, translator's note, in Hendrik, *Memoirs of Hans Hendrik*, 20; "Memoirs of Hans Hendrik, the Arctic Traveller.," *Examiner*, Nov. 16, 1878, 1465.

36. "Book Review," *Athenaeum*, 527.

37. For a brief overview of this saying and its mythical origins, see Niels Kayser Nielsen, "MYTE: Sagde Dalgas 'Hvad Udad Tabes, Skal Indad Vindes'?," Danmarks Historien, Aarhus University, http://danmarkshistorien.dk/leksikon-og-kilder/vis/materiale/myte-sagde-dalgas-hvad-udad-tabes-skal-indad-vindes/?no_cache=1 (accessed Sept. 17, 2016).

38. Kommissionen for Ledelsen af de Geologiske og Geografiske Undersøgelser i Grønland, *Meddelelser om Grønland* (Copenhagen: C. A. Reitzels Forlag, 1880), 3:xiv (hereafter *Meddelelser om Grønland*, year, and volume number).

39. The material collected was deposited at Christiansborg Castle in Copenhagen.

40. *Meddelelser om Grønland*, 1879, 1:7. Unfortunately, the collection was lost in a fire in 1884. For an overview of Danish economic imperialism in this period and in the early twentieth century, see Janina Priebe, "From Siam to Greenland: Danish Economic Imperialism at the Turn of the Twentieth Century," *Journal of World History* 27, no. 4 (2016): 619–40.

41. I draw in particular on Jørgen Sevaldsen, "'No Proper Taste for the English Way of Life': Danish Perceptions of Britain 1870–1940," in *Britain and Denmark: Political, Economic and Cultural Relations in the 19th and 20th Centuries*, ed. Jørgen Sevaldsen (Aarhus, Denmark: Museum Tusculanum Press, 2003), 68–69. For further details on the increase in internationalization of Danish research, see also Michael Bravo and Sverker Sörlin, eds., *Narrating the Arctic: A Cultural History of Nordic Scientific Practices* (Canton, MA: Science History Publications, 2002); Stefaan Blancke et al., eds., *Creationism in Europe*, Medicine, Science, and Religion in Historical Context (Baltimore: Johns Hopkins University Press, 2014).

42. Kommissionen for Ledelsen af de Geologiske og Geografiske Undersøgelser i Grønland, "Indberetning til Indenrigsministeriet om Undersøgelserne i Aarene 1878,1879 og 1880," *Meddelelser om Grønland*, 1881, 2:218.

43. Kommissionen for Ledelsen af de Geologiske og Geografiske Undersøgelser i Grønland "Indberetning til Indenrigsministeriet om Undersøgelserne i Aarene 1876, 1877 og 1878", *Meddelelser om Grønland*, 1879, 1:15.

44. Janet Martin-Nielsen, *Eismitte in the Scientific Imagination: Knowledge and Politics at the Center of Greenland* (New York: Palgrave Macmillan, 2013), 15–17. Bjarne Grønnow has written an interesting account of the interior ice in Western Greenlandic oral tradition; see Grønnow, "Blessings and Horrors of the Interior: Ethno-Historical Studies of Inuit Perceptions Concerning the Inland Region of West Greenland," *Arctic Anthropology* 46, no. 1/2 (2009): 191–201. Note that Giesecke was known as Johann Georg Metzler, Carl Ludwig Giesecke, and Charles Lewis Giesecke at various points during his career.

45. Hans Olav Thyvold, *Fridtjof Nansen: Explorer, Scientist and Diplomat*, trans. James Anderson (Norway: Font Forlag, 2012); Fridtjof Nansen, *Paa Ski over Grønland: En Skildring af den Norske Grønlands-Ekspedition 1888–89* (H. Aschehoug, 1890); David Roger Oldroyd, *Thinking about the Earth: A History of Ideas in Geology* (Cambridge, MA: Harvard University Press, 1996), 100; Rachel Laudan, *From Mineralogy to Geology: The Foundations of a Science, 1650–1830* (Chicago: University of Chicago Press, 1987).

46. A. Whittaker, "The Travels and Travails of Sir Charles Lewis Giesecke," in *Four Centuries of Geological Travel: The Search for Knowledge on Foot, Bicycle, Sledge and Camel*, ed. Patrick Wyse Jackson (London: Geological Society of London, 2007), 154.

47. Johannes Frederick Johnstrup and Hinrich Rink, eds., *Gieseckes mineralogiske rejse i Grønland* (Copenhagen: B. Lunos bogtrykkeri, 1878); Johannes Frederick Johnstrup and Japatus Steenstrup, eds., "Karl Ludwig Gieseckes Mineralogisches Reisejournal über Grönland 1806–13," in *Meddelelser om Grønland*, 1910, 35:1–478.

48. Karl Ludwig Giesecke, "Remarks on the Structure of Greenland in Support of the Opinion of Its Being an Assemblage of Islands, and Not a Continent," in *Journal of a Voyage to the Northern Whale-Fishery: Including Researches and Discoveries on the Eastern Coast of West Greenland, Made in the Summer of 1822, in the Ship Baffin of Liverpool*, by William Scoresby (Edinburgh: Archibald Constable, 1823), 467–68; Royal Geographical Society of Great Britain, ed., *Arctic Geography and Ethnology: A Selection of Papers on Arctic Geography and Ethnology. Reprinted, and Presented to the Arctic Expedition of 1875, by the President, Council, and Fellows of the Royal Geographical Society* (London: John Murray, 1875), 25.

49. Royal Geographical Society, *Arctic Geography and Ethnology*, 22, 86; Hinrich Rink, *Danish Greenland, Its People and Its Products*, ed. Robert Brown (London: H. S. King, 1877), 39.

50. *Meddelelser om Grønland*, 1879, 1:20.

51. *Meddelelser om Grønland*, 1879, 1:13.

52. In the nineteenth century glacial motion was studied by such people as James David Forbes, Louis Agassiz, John Tyndall, Thomas Henry Huxley, and Louis Rendu. See, for example, Roland Jackson, "Eunice Foote, John Tyndall and a Question of Priority," *Notes and Records: The Royal Society Journal of the History of Science*, Feb. 28, 2019, https://doi-org.ezp.lib.cam.ac.uk/10.1098/rsnr.2018.0066; Nanna Katrine Lüders Kaalund, "A Frosty Disagreement: John Tyndall, James David Forbes, and the Early Formation of the X-Club," *Annals of Science* 74, no. 4 (Oct. 2017): 282–98.

53. J. A. D. Jensen, "Vandring Paa den Grønlandske Indlandsis i Aaret 1878," *Geografisk Tidsskrift* 3 (1879): 100–107.

54. Kommissionen for Ledelsen af de Geologiske og Geografiske Undersøgelser i Grønland, "Undersøgelserne i Aarene 1878–80: Paa Vestkysten af Grönland, Indberetning til Indenrigsministeriet," *Geografisk Tidsskrift* 5 (Jan. 1881): 60.

55. Hinrich Rink, "Udsigt over Nordgrönlands Geognosi, Især med Hensyn til Bjergmassernes Mineralogiske Sammensætning," in *Om den Geographiske Beskaffenhed af de Danske Handelsdistriker i Nordgrönland, Tilligemed en Udsigt over Nordgrönlands Geognosi* (Copenhagen: B. Lunos Kgl. Hof-Bogtrykkeri, 1852), 35–62; Tobias Krüger, *Discovering the Ice Ages: International Reception and Consequences for a Historical Understanding of Climate*, first English ed., trans. Ann M. Hentschel (Leiden: Brill, [2008] 2013), 293.

56. Andreas Nicolaus Kornerup, "Geologiske Iagttagelser fra Vestkysten af Grønland" in, *Meddelelser om Grønland*, 1879, 1:80.

57. "Session 1858," in *Proceedings of the Royal Geographical Society of London* 2, no. 4 (1858): 199 (quote). The report from session 1858, which took place on April 12, 1858, also included a summary of Rink's paper; Rink, "On the Supposed Discovery of the North Coast of Greenland and an Open Polar Sea; The Great 'Humboldt Glacier,' and Other Matters Relating to the Formation of Ice in Greenland, as Described in 'Arctic Explorations in the Years 1853–4-5', by Elisha Kent Kane," in *Proceedings of the Royal Geographical Society of London* 2, no. 4 (1858): 195–97; Bache wrote a letter in response to Rink's paper; A. D. Bache, "On Dr. Rink's Remarks Respecting the Supposed Discovery by Dr. Kane of the North Coast of Greenland and an Open Polar Sea," *Proceedings of the Royal*

Geographical Society of London 2, no. 6 (1857): 359–62, 359 (quote). Rink's critique of Kane's findings were published in full as Hinrich Rink, "On the Supposed Discovery, by Dr. E. K. Kane, U. S. N., of the North Coast of Greenland, and of an Open Polar Sea, &c.; As Described in 'Arctic Explorations in the Years 1853, 1854, 1855,'" trans. Dr. Shaw, *Journal of the Royal Geographical Society of London* 28 (1858): 272–87.

58. These transnational sensitivities were discussed in the Royal Geographical Society's session for June 14, 1858, and reported by Bache, "On Dr. Rink's Remarks."

59. Hinrich Rink, "On the Supposed Discovery, by Dr. E. K. Kane," 272–73.

60. As discussed in the Royal Geographical Society's session for June 14, 1858; and John Henry Alexander, cited in *Proceedings of the Royal Geographical Society of London* 2, no. 6 (1857): 359–62.

61. "The Arctic Campaign," *Standard*, Apr. 14, 1882.

62. Georg Neumayer, "Die geographische Probleme innerhalb der Polarzonen in ihrem inneren zusammenhange Beleuchtet [Intrinsic Aspects of Geographical Problems within Polar Regions]," *Hydrographische Mittheilungen* 2, no. 5–7 (1874): 51–53; Karl Weyprecht, "Fundamental Principles of Scientific Arctic Investigation," paper presented at the annual meeting of the Academy of Science, Vienna, Jan. 18, 1875; Weyprecht, "Fundamental Principles of Arctic Investigation," paper presented at the meeting of the Association of the German Naturalists and Physicians, Graz, Sept. 18, 1875; Susan Barr and Cornelia Lüdecke, eds., *The History of the International Polar Years (IPYs)*, From Pole to Pole (New York: Springer Science & Business Media, 2010), 19; Hermann F. Koerbel, "Karl Weyprecht," in *Encyclopedia of the Arctic*, ed. Mark Nuttall (New York: Routledge, 2012), 2172–73; "Nordpolsekspeditionerne," *Jyllandsposten*, Oct. 19, 1875, 2.

63. F. W. G. Baker, "The First International Polar Year, 1882–83," *Polar Record* 21, no. 132 (1982): 277.

64. Christopher Carter, "Going Global in Polar Exploration: Nineteenth-Century American and British Nationalism and Peacetime Science," in Launius, Fleming, and DeVorkin, *Globalizing Polar Science*, 85.

65. For detailed overviews of the international collaborative efforts in the IPY, see Philip N. Cronenwett, "Publishing Arctic Science in the Nineteenth Century: The Case of the First International Polar Year," in Launius, Fleming, and DeVorkin, *Globalizing Polar Science*, 37–46; Colin P. Summerhayes, "International Collaboration in Antarctica: The International Polar Years, the International Geophysical Year, and the Scientific Committee on Antarctic Research," *Polar Record* 44, no. 4 (Oct. 2008): 321–34; Erki Tammiksaar, Natalia G. Sukhova, and Ian R. Stone, "Russia and the International Polar Year, 1882–1883," *Polar Record* 45, no. 3 (July 2009): 215–23; H. Abbes, "The German Expedition of the First International Polar Year to Cumberland Sound, Baffin Island, 1882–83," *Polar Geography and Geology* 16, no. 4 (Oct. 1992): 272–304; W. Schröder, "The First International Polar Year (1882–1883) and International Geophysical Cooperation," *Earth Sciences History* 10, no. 2 (Jan. 1991): 223–26; Rip Bulkeley, "The First Three Polar Years—A General Overview," in Barr and Luedecke, *History of the International Polar Years*, 1–6; Alexander Kraus, "Scientists and Heroes: International Arctic Cooperation at the End of the Nineteenth Century," *New Global Studies* 7, no. 2 (2013): 101–16.

66. See, for example, Jessica Ratcliff, *The Transit of Venus Enterprise in Victorian Britain* (London, Brookfield: Pickering and Chatto; Pittsburgh: University of Pittsburgh Press, 2008); Maurice Crosland, *Science under Control: The French Academy of Sciences 1795–1914* (New York: Cambridge University Press, 1992), 376–80; Rothenberg, "Making Science Global?," 28.

67. Shelagh D. Grant, *Polar Imperative: A History of Arctic Sovereignty in North America* (Vancouver: Douglas & McIntyre, 2010), 95.

68. Trevor H. Levere, *Science and the Canadian Arctic: A Century of Exploration, 1818–1918* (Cambridge: Cambridge University Press, 2004), 322–33.

69. "The Circumpolar Stations," *Times*, Aug. 16, 1883, 7; "Arctic Campaign," 6.

70. "Log Book," *Geographical Magazine*, Apr. 1, 1876, 105.

71. Clements Robert Markham to Robert Brown of Campster, London, Dec. 9, 1876, MS 441/9/21, SPRI.

72. See, for example, Clements Robert Markham to Robert Brown, Nov. 18, 1875, MS 441/9/13; Markham to Brown, Dec. 28, 1875, MS 441/9/15, both in SPRI.

73. Clements Robert Markham, "Multiple News Items, Arctic Exploration," *Standard*, Dec. 16, 1880; George Richards, "Multiple News Items, Arctic Exploration," *Standard*, Dec. 15, 1880. See also "The Arctic Campaign of 1882–3," *Times*, Jan. 19, 1883, 3; "The Royal Society," *Nature* 27 (Dec. 14, 1882): 162.

74. Clements Robert Markham to Robert Brown, Dec. 15, 1880, MS 441/9/22, SPRI; "Multiple News Items," *Standard*, May 23, 1882.

75. The Great Slave Lake is on the border between the sub-Arctic and the Arctic

76. "Science Notices," *Dublin Review, 1836–1910; London*, Apr. 1883, 463–64.

77. Robert E. Kohler, *Landscapes and Labscapes: Exploring the Lab-Field Border in Biology* (Chicago: University of Chicago Press, 2002), 3. See also Henrika Kuklick and Robert E. Kohler, "Introduction," *Osiris* 11 (1996): 1–14.

78. John Ross, *A Voyage of Discovery, Made under the Orders of the Admiralty, in His Majesty's Ships Isabella and Alexander, for the Purpose of Exploring Baffin's Bay, and Inquiring into the Probability of a North-West Passage* (London: John Murray, 1819), 9.

79. Karen M. Morin, *Civic Discipline: Geography in America, 1860–1890* (New York: Routledge, 2016), 141; R. G. Barry, "Climate: Research Programs," in *Encyclopedia of the Arctic*, ed. Mark Nuttall (New York: Routledge, 2012), 379; Barr and Lüdecke, *History of the International Polar Years*, 3.

80. Henry P. Dawson, "Report on the Circumpolar Expedition to Fort Rae," *Proceedings of the Royal Society of London* 36, no. 228–31 (1883): 173–79; Henry P. Dawson, *Observations of the International Polar Expeditions, 1882–83: Fort Rae* (London: Eyre and Spottiswood for Trübner, 1886). The report was also published as "Observations of the International Polar Expeditions," *Nature* 35, no. 147 (Dec. 1886): 147. For a detailed summary of Denmark's contribution to the First IPY, see Barr and Lüdecke, *History of the International Polar Years (IPYs)*, 40–42. For an overview of the scientific results of the Fort Rae station, see Yong Zhou, *The Histories of the International Polar Years and the Inception and Development of the International Geophysical Year*, 1st ed., Annals of the International Geophysical Year (London: Pergamon, 1959), 1:26–29.

81. Adolphus W. Greely, *Three Years of Arctic Service; an Account of the Lady Franklin Bay Expedition of 1881–84, and the Attainment of the Farthest North* (New York: C. Scribner's Sons, 1886), 1:preface. See also George Lippard Barclay, *The Greely Arctic Expedition as Fully Narrated by Lieut. Greely, U.S.A., and Other Survivors: Full Account of the Terrible Sufferings on the Ice* (Philadelphia: Barclay, 1884); and George Melville, *In the Lena Delta: A Narrative of the Search for Lieut-Commander De Long and His Companions, Followed by an Account of the Greely Relief Expedition and a Proposed Method of Reaching the North Pole* (Boston: Houghton Mifflin, 1884).

82. Henry P. Dawson, *Observations of the International Polar Expeditions*; Adolphus W. Greely, *International Polar Expedition: Report on the Proceedings of the United States Expedition to Lady Franklin Bay, Grinnell Land* (Washington, DC: US Government Publishing Office, 1888).

83. Dawson, *Observations of the International Polar Expeditions*, vii–viii.

84. Dawson, *Observations of the International Polar Expeditions*, xi.

85. Translation is from Baker, "First International Polar Year," 277; Weyprecht, "Fundamental Principles of Scientific Arctic Investigation"; Weyprecht, "Fundamental Principles of Arctic Investigation."

86. Kohler, *Landscapes and Labscapes*, 5; David N. Livingstone, *Putting Science in Its Place: Geographies of Scientific Knowledge* (Chicago: University of Chicago Press, 2010), 47.

87. Debra J. Lindsay, *Science in the Subarctic: Trappers, Traders, and the Smithsonian Institution* (Washington, DC: Smithsonian Institution Press, 1993).

88. John Knox Laughton, "An Address Delivered at the Annual General Meeting, January 16th, 1884," *Quarterly Journal of the Royal Meteorological Society* 10, no. 50 (Apr. 1884): 82.

89. For a general overview of Peary and his construction of the heroic Arctic explorer image, see Lyle Dick, "Aboriginal-European Relations during the Great Age of North Polar Exploration," *Polar Geography* 26, no. 1 (Jan. 2002): 66–86; Bruce Henderson, *True North: Peary, Cook, and the Race to the Pole* (New York: W. W. Norton, 2006); Lyle Dick, "Robert Peary's North Polar Narratives and the Making of an American Icon," *American Studies* 45, no. 2 (2004): 5–34; Kelly Lankford, "Arctic Explorer Robert E. Peary's Other Quest: Money, Science, and the Year 1897," *American Nineteenth Century History* 9, no. 1 (Mar. 2008): 37–60; Douglas W. Wamsley, "'We Are Fully in the Expedition': Philadelphia's Support for the North Greenland Expeditions of Robert E. Peary, 1891–1895," *Geographical Review* 107, no. 1 (Jan. 2017): 207–35; Genevieve LeMoine, "Elatu's Funeral: A Glimpse of Inughuit-American Relations on Robert E. Peary's 1898–1902 Expedition," *Arctic* 67, no. 3 (2014): 340–46; Patricia A. M. Huntington, "Robert E. Peary and the Cape York Meteorites," *Polar Geography* 26, no. 1 (Jan. 2002): 53–65; Baron Bedesky, *Peary and Henson: The Race to the North Pole* (New York: Crabtree, 2006); A. C. Bonga, "Robert E. Peary: A Medical Assessment," *Polar Record* 28, no. 164 (Jan. 1992): 71–72; Bonga, "Robert E. Peary and Bob Bartlett: A Rejoinder from the Author," *Polar Record* 28, no. 166 (July 1992): 252; Henderson, *True North*.

90. John Ambrose Fleming, "The Proposed Second International Polar Year, 1932–1933," *Geographical Review* 22, no. 1 (1932): 131.

Epilogue

1. For a detailed account of Peary's North Pole venture, see, for example, Bruce Henderson, *True North: Peary, Cook, and the Race to the Pole* (New York: W. W. Norton, 2006); Fergus Fleming, *Ninety Degrees North: The Quest for the North Pole* (New York: Grove Press, 2007); Daniel E. Harmon, *Robert Peary and the Quest for the North Pole* (New York: Infobase Publishing, 2001).

2. Walter Wellman, "Wellman Riddles Cook's Narrative: Concludes, after Analysis, That His Arctic Journey and Report Were Planned Beforehand," *New York Times*, Nov. 29, 1909, 1.

3. Henrika Kuklick, "Personal Equations: Reflections on the History of Fieldwork, with Special Reference to Sociocultural Anthropology," *Isis* 102, no. 1 (2011): 1–33.

BIBLIOGRAPHY

Archives
National Archives, Washington, DC
Royal Society Archives, London
Scott Polar Research Institute, University of Cambridge
Statsbiblioteket, Aarhus Universitet, Denmark

Books, Journals, Magazines
Abbes, H. "The German Expedition of the First International Polar Year to Cumberland Sound, Baffin Island, 1882–83." *Polar Geography and Geology* 16, no. 4 (October 1992): 272–304.
Adele, Perry. "Designing Dispossession: The Select Committee on the Hudson's Bay Company, Fur-Trade Governance, Indigenous Peoples and Settler Possibility." In *Indigenous Communities and Settler Colonialism: Land Holding, Loss and Survival in an Interconnected World*, edited by Zoë Laidlaw and Alan Lester, 158–72. New York: Palgrave Macmillan, 2015.
Andersen, Casper, Jakob Bek-Thomsen, and Peter C. Kjærgaard. "The Money Trail: A New Historiography for Networks, Patronage, and Scientific Careers." *Isis* 103, no. 2 (2012): 310–15.
Andersen, Casper, and Hans H. Hjermitslev. "Directing Public Interest: Danish Newspaper Science 1900–1903." *Centaurus* 51, no. 2 (May 2009): 143–67.
Andersen, Preben. "Herrnhutterne i Grønland." *Tidsskriftet Grønland*, no. 2 (1969): 50–64.
Anderson, Katharine. *Predicting the Weather: Victorians and the Science of Meteorology*. Chicago: University of Chicago Press, 2005.
Andra-Warner, Elle. *Hudson's Bay Company Adventures: Tales of Canada's Fur Traders*. 2nd ed. Victoria, BC: Heritage House, 2003.
Andrewes, William J. H. ed. *The Quest for Longitude: The Proceedings of the Longitude Symposium, Harvard University, Cambridge, Massachusetts, November 4–6, 1993*. Cambridge, MA: Collection of Historical Scientific Instruments, Harvard University, 1996.
Anonymous Missionary. "Udtog af en Dansk Dames Dagbog, Ført i Grønland 1837–1838." Part 1. *Læsefrugter*, January 1839, 105–7.
Anonymous Missionary. "Udtog af en Dansk Dames Dagbog, Ført i Grønland 1837–1838." Part 2. *Læsefrugter*, February 1839, 231–34.
Apollonio, Spencer. *Lands That Hold One Spellbound: A Story of East Greenland*. Calgary: University of Calgary Press, 2008.
"Arctic Searching Expedition: A Journal of a Boat-Voyage through Rupert's Land and the Arctic Sea, in Search of the Discovery Ships under Command of Sir John Franklin. With an Appendix on the Physical Geography of North America." *Athenaeum*, November 29, 1851, 1246–47.
"Arctic Searching Expedition: A Journal of a Boat Voyage through Rupert's Land and the Arctic Sea, in Search of the Discovery Ships under Command of Sir John Franklin. With an Appendix on the Physical Geography of North America." *North British Review*, February 1852, 445–89.

Armour, David A. "Biography—Dease, John." *Dictionary of Canadian Biography*. Vol. 5, *1801–1820*. Accessed November 2, 2016. http://www.biographi.ca/en/bio/dease_john_5E.html.

Arnold Arboretum. *Sargentia: A Continuation of the Contributions from the Arnold Arboretum of Harvard University*. Cambridge, MA: Arnold Arboretum of Harvard University, 1943.

"ART. V.—Narrative of the Discoveries on the North Coast of America, Effected by the Officers of the Hudson's Bay Company, during the Years 1836–39." *Quarterly Review*, December 1843, 113–29.

"ART. VIII. 1. A Voyage of Discovery, Made, under the Orders of the Admiralty, in H. M. Ships Isabella and Alexander, for the Purpose of Exploring Baffin's Bay, and Inquiring into the Probability of a North-West Passage." *Edinburgh Monthly Review*, June 1819, 726–46.

"ART. XIX.—A Voyage of Discovery Made under the Orders of the Admiralty in His Majesty's Ships Isabella and Alexander, for the Purpose of Exploring Baffin's Bay, and Inquiring into the Probability of a North-West Passage." *British Review, and London Critical Journal, 1811–1825*, May 1819, 413–39.

Auring, Steffen. *Dansk litteraturhistorie*. Vol. 5, *Borgerlig enhedskultur 1807–48*. Copenhagen: Gyldendal, 1984.

Austin, Alvyn, and Jamie S. Scott, eds. *Canadian Missionaries, Indigenous Peoples: Representing Religion at Home and Abroad*. Toronto: University of Toronto Press, 2005.

Bache, A. D. "On Dr. Rink's Remarks Respecting the Supposed Discovery by Dr. Kane of the North Coast of Greenland and an Open Polar Sea." *Proceedings of the Royal Geographical Society of London* 2, no. 6 (1857): 359–62.

Baker, Alexi. "Longitude Essays." Cambridge Digital Library. Accessed March 21, 2016. http://cudl.lib.cam.ac.uk/view/ES-LON-00023/1.

Baker, F. W. G. "The First International Polar Year, 1882–83." *Polar Record* 21, no. 132 (1982): 275–85.

Bang, Peter Fibiger, and Dariusz Kolodziejczyk. *Universal Empire: A Comparative Approach to Imperial Culture and Representation in Eurasian History*. Cambridge: Cambridge University Press, 2012.

Barclay, George Lippard. *The Greely Arctic Expedition as Fully Narrated by Lieut. Greely, U.S.A., and Other Survivors: Full Account of the Terrible Sufferings on the Ice*. Philadelphia: Barclay, 1884.

Barr, Susan, and Cornelia Lüdecke, eds. *The History of the International Polar Years (IPYs)*. New York: Springer Science & Business Media, 2010.

Barr, William, ed. *From Barrow to Boothia: The Arctic Journal of Chief Factor Peter Warren Dease, 1836–1839*. Montreal: McGill-Queen's University Press, 2002.

Barr, William. "Searching for Franklin Where He Was Ordered to Go: Captain Erasmus Ommanney's Sledging Campaign to Cape Walker and Beyond, Spring 1851." *Polar Record* 52, no. 4 (July 2016): 474–98.

Barr, William. "The Use of Dog Sledges during the British Search for the Missing Franklin Expedition in the North American Arctic Islands, 1848–59." *Arctic* 62, no. 3 (2009): 257–72.

[Barrow, John]. "ART. I.-1. Narrative of a Second Voyage in Search of a Northwest Passage, and of a Residence in the Arctic Regions, during the Years 1829–30–31–32–33." *Quarterly Review*, July 1835, 1–39.

Barrow, John. *Voyages of Discovery and Research within the Arctic Regions, from the Year 1818 to the Present Time: Under the Command of the Several Naval Officers Employed by Sea and Land in Search of a Northwest Passage from the Atlantic to the Pacific; with Two Attempts to Reach the North Pole. Abridged and Arranged from the Official Narratives, with Occasional Remarks*. London: John Murray, 1846.

Barry, R. G. "Climate: Research Programs." In *Encyclopedia of the Arctic*, edited by Mark Nuttall, 379–84. New York: Routledge, 2012.

Baumgart, Winfried. *Imperialism: The Idea and Reality of British and French Colonial Expansion.* Oxford: Oxford University Press, 1982.

Bayly, C. A., Sven Beckert, Matthew Connelly, Isabel Hofmeyr, Wendy Kozol, and Patricia Seed. "On Transnational History." AHR Conversation. *American Historical Review* 111, no. 5 (December 2006): 1441–64.

Bedesky, Baron. *Peary and Henson: The Race to the North Pole.* New York: Crabtree, 2006.

Beechey, Frederick William. *A Voyage of Discovery towards the North Pole: Performed in His Majesty's Ships Dorothea and Trent, under the Command of Captain David Buchan, R.N.; 1818; to Which Is Added, a Summary of All the Early Attempts to Reach the Pacific by Way of the Pole.* London: R. Bentley, 1843.

Beer, Gillian. *Open Fields: Science in Cultural Encounter.* Oxford: Oxford University Press, 1999.

Belknap, Geoffrey. *From a Photograph: Authenticity, Science and the Periodical Press, 1870–1890.* London: Bloomsbury, 2016.

Bella, Elizabeth. "British Arctic Expedition." In *Antarctica and the Arctic Circle: A Geographic Encyclopedia of the Earth's Polar Regions*, edited by Andrew Jon Hund, 161–62. Santa Barbara, CA: ABC-CLIO, 2014.

Bevir, Mark, ed. *Historicism and the Human Sciences in Victorian Britain.* Cambridge: Cambridge University Press, 2017.

Binnema, Ted. *Enlightened Zeal: The Hudson's Bay Company and Scientific Networks, 1670–1870.* Buffalo, NY: University of Toronto Press, 2014.

Blancke, Stefaan, Hans Henrik Hjermitslev, and Peter C. Kjærgaard, eds. *Creationism in Europe, Medicine, Science, and Religion in Historical Context.* Baltimore: Johns Hopkins University Press, 2014.

Bleichmar, Daniela. *Visible Empire: Botanical Expeditions and Visual Culture in the Hispanic Enlightenment.* Chicago: University of Chicago Press, 2012.

Boehmer, Elleke, and Bart Moore-Gilbert. "Introduction to Special Issue: Postcolonial Studies and Transnational Resistance." *Interventions* 4, no. 1 (January 2002): 7–21.

Bohls, Elizabeth A., and Ian Duncan, eds. *Travel Writing 1700–1830: An Anthology.* Oxford World's Classics. Oxford: Oxford University Press, 2008.

Bonga, A. C. "Robert E. Peary: A Medical Assessment." *Polar Record* 28, no. 164 (January 1992): 71–72.

Bonga, A. C. "Robert E. Peary and Bob Bartlett: A Rejoinder from the Author." *Polar Record* 28, no. 166 (July 1992): 252.

"Book Review." *Athenaeum*, October 26, 1878, 527–28.

"Book Review." *Monthly Review*, September 1843, 76–85.

Bose, Sugata. *A Hundred Horizons: The Indian Ocean in the Age of Global Empire.* Cambridge, MA: Harvard University Press, 2006.

Braithwaite, John. *Supplement to Captain Sir John Ross's Narrative of a Second Voyage in the Victory, in Search of a North-West Passage. Containing the Suppressed Facts Necessary to a Proper Understanding of the Causes of the Failure of the Steam Machinery of the Victory, and a Just Appreciation of Captain Sir John Ross's Character as an Officer and a Man of Science.* London: Chapman & Hall, 1835.

Brandt, Anthony. *The Man Who Ate His Boots: Sir John Franklin and the Tragic History of the Northwest Passage.* New York: Random House, 2011.

Bravo, Michael Trevor. "Ethnographic Navigation and the Geographical Gift." In *Geography and Enlightenment*, edited by David N. Livingstone and Charles W. J. Withers, 199–235. Chicago: University of Chicago Press, 1999.

Bravo, Michael Trevor. "The Postcolonial Arctic." *Moving Worlds: A Journal of Transcultural Writings* 15 (2015): 93–111.

Bravo, Michael, and Sverker Sörlin, eds. *Narrating the Arctic: A Cultural History of Nordic Scientific Practices*. Canton, MA: Science History Publications, 2002.

[Brewster, David]. "ART. VII.—Narrative of a Second Voyage in Search of a Northwest Passage, and of a Residence in the Arctic Regions during the Years 1829, 1830, 1831, 1832, 1833," *Edinburgh Review, 1802–1929*, July 1835, 417–53.

British Association for the Advancement of Science. *Report of the 17th Meeting of the British Association for the Advancement of Science (Oxford)*. London: Taylor & Francis, 1848.

Browne, Janet. "Biogeography and Empire." In *Cultures of Natural History*, edited by Nicholas Jardine, James A. Secord, and E. C. Spary, 305–21. Cambridge: Cambridge University Press, 1996.

Browne, Janet. *The Secular Ark: Studies in the History of Biogeography*. New Haven, CT: Yale University Press, 1983.

Bryld, Tine. *I den Bedste Mening*. Copenhagen: Gyldendal, 2010.

Bulkeley, Rip. "The First Three Polar Years—A General Overview." In *The History of the International Polar Years (IPYs)*, edited by Susan Barr and Cornelia Luedecke, 1–6. From Pole to Pole. Berlin: Springer, 2010.

Burbank, Jane, and Frederick Cooper. *Empires in World History: Power and the Politics of Difference*. Princeton, NJ: Princeton University Press, 2010.

Burley, Edith. *Servants of the Honourable Company: Work, Discipline, and Conflict in the Hudson's Bay Company, 1770–1870*. Toronto: Oxford University Press, 1997.

"By Land and Sea." *Fraser's Magazine for Town and Country, 1830–1869; London*, September 1853, [251]–64.

Cabañas, Miguel A., Jeanne Dubino, Veronica Salles-Reese, and Gary Totten, eds. *Politics, Identity, and Mobility in Travel Writing*. New York: Routledge, 2015.

Cantor, Geoffrey, Gowan Dawson, Richard Noakes, Sally Shuttleworth, and Jonathan Topham. *Science in the Nineteenth-Century Periodical: Reading the Magazine of Nature*. Cambridge: Cambridge University Press, 2004.

"Captain Ross, and Sir James Lancaster's Sound." *Blackwood's Edinburgh Magazine*, May 1819, 150–51.

Carter, Christopher. "Going Global in Polar Exploration: Nineteenth-Century American and British Nationalism and Peacetime Science." In *Globalizing Polar Science: Reconsidering the International Polar and Geophysical Years*, edited by Roger D. Launius, James Rodger Fleming, and David H. DeVorkin, 85–106. Palgrave Studies in the History of Science and Technology. New York: Palgrave Macmillan, 2010.

Cavell, Janice. "Going Native in the North: Reconsidering British Attitudes during the Franklin Search, 1848–1859." *Polar Record* 45, no. 1 (January 2009): 25–35.

Cavell, Janice. "Publishing Sir John Franklin's Fate: Cannibalism, Journalism, and the 1881 Edition of Leopold McClintock's The Voyage of the 'Fox' in the Arctic Seas." *Book History* 16, no. 1 (October 2013): 155–84.

Cavell, Janice. *Tracing the Connected Narrative: Arctic Exploration in British Print Culture, 1818–1860*. Toronto: University of Toronto Press, 2008.

Cawood, John. "Terrestrial Magnetism and the Development of International Collaboration in the Early Nineteenth Century." *Annals of Science* 34, no. 6 (November 1977): 551–87.

Chapin, David. *Exploring Other Worlds: Margaret Fox, Elisha Kent Kane, and the Antebellum Culture of Curiosity*. Amherst: University of Massachusetts Press, 2004.

Christensen, Dan Ch. *Hans Christian Ørsted: Reading Nature's Mind*. Oxford: Oxford University Press, 2013.

Clifford, James. *Routes: Travel and Translation in the Late Twentieth*. Cambridge, MA: Harvard University Press, 1999.

Coates, Colin. "Like 'The Thames towards Putney': The Appropriation of Landscape in Lower Canada." *Canadian Historical Review* 74, no. 3 (September 1993): 317–43.

Cookman, Scott. *Ice Blink: The Tragic Fate of Sir John Franklin's Lost Polar Expedition*. New York: John Wiley & Sons, 2001.

Craciun, Adriana. "Writing the Disaster: Franklin and Frankenstein." *Nineteenth-Century Literature* 65, no. 4 (March 2011): 433–80.

Cronenwett, Philip N. "British Arctic Expedition, 1875–1876." In *Encyclopedia of the Arctic*, edited by Mark Nuttall, 277–78. New York: Routledge, 2012.

Cronenwett, Philip N. "Publishing Arctic Science in the Nineteenth Century: The Case of the First International Polar Year." In *Globalizing Polar Science*, edited by Roger D. Launius, James Rodger Fleming, and David H. DeVorkin, 37–46. Palgrave Studies in the History of Science and Technology. New York: Palgrave Macmillan, 2010.

Crosland, Maurice. *Science under Control: The French Academy of Sciences 1795–1914*. New York: Cambridge University Press, 1992.

Crowe, Keith J. *A History of the Original Peoples of Northern Canada*. Rev. ed. Montreal: McGill-Queen's University Press, [1974] 1991.

Cruikshank, Julie. *Do Glaciers Listen? Local Knowledge, Colonial Encounters, and Social Imagination*. Vancouver: UBC Press, 2010.

Cruwys, Liz. "Henry Grinnell and the American Franklin Searches." *Polar Record* 26, no. 158 (July 1990): 211–16.

Daston, Lorraine, and Peter Galison. *Objectivity*. New York: Zone Books, 2007.

David, Robert G. *The Arctic in the British Imagination 1818–1914*. Manchester: Manchester University Press, 2000.

Davis, Richard Clarke. *Lobsticks and Stone Cairns: Human Landmarks in the Arctic*. Calgary: University of Calgary Press, 1996.

Dawson, Henry P. "Observations of the International Polar Expeditions," *Nature* 35, no. 147 (December 1886): 147.

Dawson, Henry P. *Observations of the International Polar Expeditions, 1882–83: Fort Rae*. London: Eyre and Spottiswood for Trübner, 1886.

Dawson, Henry P. "Report on the Circumpolar Expedition to Fort Rae." *Proceedings of the Royal Society of London* 36, no. 228–31 (1883): 173–79.

Desmond, Adrian, and James Moore. *Darwin's Sacred Cause: Race, Slavery and the Quest for Human Origins*. London: Penguin Books, 2009.

Dettelbach, Michael. "The Face of Nature: Precise Measurement, Mapping, and Sensibility in the Work of Alexander von Humboldt." *Studies in History and Philosophy of Science Part C: Studies in History and Philosophy of Biological and Biomedical Sciences* 30, no. 4 (1999): 473–504.

Diala, Isidore. "Colonial Mimicry and Postcolonial Re-membering in Isidore Okpewho's Call Me by My Rightful Name." *Journal of Modern Literature* 36, no. 4 (2013): 77–95.

Dick, Lyle. "Aboriginal-European Relations during the Great Age of North Polar Exploration." *Polar Geography* 26, no. 1 (January 2002): 66–86.

Dick, Lyle. "Robert Peary's North Polar Narratives and the Making of an American Icon." *American Studies* 45, no. 2 (2004): 5–34.

Dickens, Charles. "The Lost Arctic Voyagers." Part 1. *Household Words*, December 2, 1854, 361–65.

Dickens, Charles. "The Lost Arctic Voyagers." Part 2. *Household Words*, December 9, 1854, 385–93.

"Discoveries on the North Coast of America." *Chambers' Edinburgh Journal*, September 1844, 277–78.

Distad, Merrill. "Newspapers and Magazines." In *History of the Book in Canada: 1840–1918*, edited by Patricia Fleming, Yvan Lamonde, and Fiona Black, 2:293–302. Toronto: University of Toronto Press, 2005.

Drummond, Thomas. "Sketch of a Journey to the Rocky Mountains and to the Columbia River in North America," In *Botanical Miscellany*, ed. William Hooker, 1:178–219. London: John Murray, 1830.

Edwards, Paul N. *A Vast Machine: Computer Models, Climate Data, and the Politics of Global Warming.* Cambridge, MA: MIT Press, 2010.

Egede, Hans. *Det Gamle Grønlands nye Perlustration: Eller Naturel-Historie, og Beskrivelse over det Gamle Grønlands Situation, Luft, Temperament og Beskaffenhed.* Copenhagen: J. C. Groth, 1741.

Eiselein, Gregory. *Literature and Humanitarian Reform in the Civil War Era.* Bloomington: Indiana University Press, 1996.

Erslew, Thomas Hansen. *Almindeligt Forfatter-Lexicon for Kongeriget Danmark med Tilhørende Bilande, fra 1814 til 1840.* Copenhagen: Forlagsforeningens Forlag, 1843.

Fanon, Frantz. *The Wretched of the Earth.* Translated by Richard Philcox. New York: Grove Press, [1961] 2007.

Feldbæk, Ole. *Danmarks Historie.* Copenhagen: Gyldendal, 2010.

Fetherling, George. *The Rise of the Canadian Newspaper.* Toronto: Oxford University Press, 1990.

Fischer-Tiné, Harald, and Michael Mann, eds. *Colonialism as Civilizing Mission: Cultural Ideology in British India.* London: Anthem Press, 2004.

Fjagesund, Peter. *The Dream of the North: A Cultural History to 1920.* Amsterdam: Rodopi, 2014.

Fleming, Fergus. *Barrow's Boys.* New York: Atlantic Monthly Press, 2000.

Fleming, Fergus. *Ninety Degrees North: The Quest for the North Pole.* New York: Grove Press, 2007.

Fleming, J. A. "The Proposed Second International Polar Year, 1932–1933." *Geographical Review* 22, no. 1 (1932): 131–34.

Fleming, James, and Vladimir Jankovic. "Revisiting Klima." *Osiris* 26, no. 1 (2011): 1–15.

Fleming, Patricia, Yvan Lamonde, and Fiona Black, eds. *History of the Book in Canada.* Vol. 2, *1840–1918.* Toronto: University of Toronto Press, 2005.

Fleming, Patricia, Yvan Lamonde, and Giles Gallichan, eds. *History of the Book in Canada.* Vol. 1, *Beginnings to 1840.* Toronto: University of Toronto Press, 2004.

Flora Danica. Det Kongelige Bibliotek. 51 vols. Accessed December 11, 2015. http://www.kb.dk/da/materialer/kulturarv/institutioner/DetKongeligeBibliotek/Billeder_oversigt/flora_danica.html.

[Francis, Egerton]. "ART. V.—A Voyage of Discovery, Made under the Orders of the Admiralty, in His Majesty's Ships Isabella and Alexander, for the Purpose of Exploring Baffin's Bay, and Inquiring into the Probability of a North-West Passage." *Edinburgh Review, 1802–1929*, March 1819, 336–68.

Francis, Egerton. *Essays on History, Biography, Geography, Engineering &c. Contributed to the Quarterly Review by the Late Earl of Ellesmere.* London: John Murray, 1858.

Franklin, John. *Narrative of a Journey to the Shores of the Polar Sea in the Years 1819, 20, 21 and 22, with an Appendix on Various Subjects Relating to Science and Natural History Illustrated by Numerous Plates and Maps.* London: John Murray, 1823.

Franklin, John. *Narrative of a Second Expedition to the Shores of the Polar Sea, in the Year 1825, 1826 and 1827: Including an Account of the Progress of a Detachment to the Eastward by John Richardson; Illustrated by Numerous Plates and Maps. Published by Authority of the Right Honourable the Secretary of State for Colonial Affairs.* London: John Murray, 1828.

Funch, Johan Christian Wilhelm. *Syv aar i Nordgrönland.* Viborg, Denmark: Rabell, 1840.

Fyfe, Aileen. *Science and Salvation: Evangelical Popular Science Publishing in Victorian Britain.* Chicago: University of Chicago Press, 2004.

Fyfe, Aileen. *Steam-Powered Knowledge: William Chambers and the Business of Publishing, 1820–1860.* Chicago: University of Chicago Press, 2012.

Fyfe, Aileen, and Bernard Lightman. *Science in the Marketplace: Nineteenth-Century Sites and Experiences.* Chicago: University of Chicago Press, 2007.

Garboe, Axel. *Geologiens Historie i Danmark.* Vol. 2, *Forskere og Resultater.* Copenhagen: C. A. Reitzel, 1961.

Garrison, Laurie. "Virtual Reality and Subjective Responses: Narrating the Search for the Franklin Expedition through Robert Burford's Panorama." *Early Popular Visual Culture* 10, no. 1 (2012): 7–22.

Geiger, John, and Owen Beattie. *Frozen in Time.* London: Bloomsbury, [1987] 2012.

Geiger, John, and Alanna Mitchell. *Franklin's Lost Ship: The Historic Discovery of HMS Erebus.* Toronto: HarperCollins, 2015.

Gillespie, Beryl. "Yellowknife." In *Handbook of North American Indians: Subarctic,* edited by William C. Sturtevant, 285–90. Washington, DC: Government Printing Office, 1978.

Glenthøj, Rasmus, and Morten Nordhagen Ottosen. *Experiences of War and Nationality in Denmark and Norway, 1807–1815.* London: Palgrave Macmillan UK, 2014.

Goldstein, Robert Justin. *Political Censorship of the Arts and the Press in Nineteenth-Century Europe.* New York: St. Martin's Press, 1989.

Gooday, Graeme. *The Morals of Measurement: Accuracy, Irony, and Trust in Late Victorian Electrical Practice.* Cambridge: Cambridge University Press, 2004.

Gorham, Harriet. "Tattannoeuck (Augustus)." *The Canadian Encyclopedia.* Accessed January 31, 2017. http://www.thecanadianencyclopedia.ca/en/article/augustus/.

Gorman, Daniel. *Imperial Citizenship: Empire and the Question of Belonging.* Manchester: Manchester University Press, 2010.

Graah, Wilhelm August. *Narrative of an Expedition to the East Coast of Greenland, Sent by Order of the King of Denmark, in Search of the Lost Colonies.* Translated by G. Gordon Macdougall. First English ed. London: Royal Geographical Society of London, 1837.

Graah, Wilhelm August. *Undersøgelses-Reise til Østkysten af Grønland: Efter Kongelign Befaling Udført i Aarene 1828–31.* Copenhagen: J. D. Qvist, 1832.

Grace, Sherrill E. "Gendering Northern Narrative." In *Echoing Silence: Essays on Arctic Narrative,* edited by John George Moss, 163–83. Ottawa: University of Ottawa Press, 1997.

Grant, Shelagh D. *Polar Imperative: A History of Arctic Sovereignty in North America.* Vancouver: Douglas & McIntyre, 2010.

Great Britain. Royal Navy. Admiralty. *Arctic Expedition: Papers and Correspondence Relating to the Equipment and Fitting Out of the Arctic Expedition of 1875, Including Report of the Admiralty Arctic Committee. Presented to Both Houses of Parliament by Command of Her Majesty.* London: George Edward Eyre and William Spottiswoode, 1875.

Greely, Adolphus W. *International Polar Expedition: Report on the Proceedings of the United States Expedition to Lady Franklin Bay, Grinnell Land*. Washington, DC: Government Publishing Office, 1888.

Greely, Adolphus W. *Three Years of Arctic Service; An Account of the Lady Franklin Bay Expedition of 1881–84, and the Attainment of the Farthest North*. 2 vols. New York: C. Scribner's Sons, 1886.

Greene, Mott T. *Geology in the Nineteenth Century: Changing Views of a Changing World*. Ithaca, NY: Cornell University Press, 1982.

Groden, Michael, Martin Kreiswirth, and Imre Szeman, eds. *Contemporary Literary and Cultural Theory: The Johns Hopkins Guide*. Baltimore: Johns Hopkins University Press, 2012.

Grønnow, Bjarne. "Blessings and Horrors of the Interior: Ethno-Historical Studies of Inuit Perceptions Concerning the Inland Region of West Greenland." *Arctic Anthropology* 46, no. 1/2 (2009): 191–201.

Grove, Richard H. *Green Imperialism: Colonial Expansion, Tropical Island Edens and the Origins of Environmentalism, 1600–1860*. Cambridge: Cambridge University Press, 1995.

Hall, Catherine. *Civilising Subjects: Metropole and Colony in the English Imagination 1830–1867*. Chicago: University of Chicago Press, 2002.

Hansen, Klaus Georg. "Wilhelm August Graah." In *Encyclopedia of the Arctic*, edited by Mark Nuttall, 763–64. New York: Routledge, 2012.

Harmon, Daniel E. *Robert Peary and the Quest for the North Pole*. New York: Infobase, 2001.

Harrison, Phyllis. *The Home Children: Their Personal Stories*. Winnipeg: Watson and Dwyer, 1979.

Hastrup, Kirsten. "Anticipating Nature: The Productive Uncertainty of Climate Models." In *The Social Life of Climate Change Models: Anticipating Nature*, edited by Kirsten Hastrup and Martin Skrydstrup, 1–29. New York: Routledge, 2012.

Headrick, Daniel R. *Power over Peoples: Technology, Environments, and Western Imperialism, 1400 to the Present*. Princeton, NJ: Princeton University Press, 2012.

Helm, June, Teresa S. Carterette, and Nancy Oestreich Lurie. *The People of Denendeh: Ethnohistory of the Indians of Canada's Northwest Territories*. Iowa City: University of Iowa Press, 2000.

Henderson, Bruce B. *Fatal North: Adventure and Survival aboard USS Polaris, the First U.S. Expedition to the North Pole*. New York: New American Library, 2001.

Henderson, Bruce. *True North: Peary, Cook, and the Race to the Pole*. New York: W. W. Norton, 2006.

Hendrik, Hans [Suersaq]. *Memoirs of Hans Hendrik: The Arctic Traveller, Serving under Kane, Hayes, Hall and Nares, 1853–1876*. Edited by George Stephens. Translated by Hinrich Rink. London: Trübner, 1878.

Hevly, Bruce. "The Heroic Science of Glacier Motion." *Osiris* 11 (January 1996): 66–86.

Hewitt, Martin. *The Dawn of the Cheap Press in Victorian Britain: The End of the "Taxes on Knowledge," 1849–1869*. London: Bloomsbury, 2014.

Hewson, J. B. *A History of the Practice of Navigation*. Rev. ed. Glasgow: Brown, Son & Ferguson, [1951] 1983.

Hill, Jen. *White Horizon: The Arctic in the Nineteenth-Century British Imagination*. Albany: State University of New York Press, 2009.

Hjermitslev, Hans Henrik. "Naturvidenskabens Rolle På de Danske Folkehøjskoler." In *Två Sidor af Samma Mynt? Folkbilding och Yrkesutbildning Vid de Nordiska Folkhögskolorna*, ed. Fay L. Nilsson and Anders Nilsson, 111–38. Lund, Sweden: Nordic Academic Press, 2010.

Hooker, William Jackson, ed. *Flora Boreali-Americana, or, The botany of the northern parts of British America, compiled principally from the plants collected by Dr. Richardson & Mr. Drummond on the late northern expeditions, under command of Captain Sir John Franklin, R.N. to which are added (by permission of the Horticultural Society of London) those of Mr. Douglas from north west America and other naturalists.* 2 vols. London: Henry G. Bohn, [1829]–40.

Horden, P., and N. Purcell. "The Mediterranean and 'the New Thalassology.'" *American Historical Review* 111, no. 3 (June 2006): 722–40.

Hørring, Hugo. *Bemærkninger til Justitsraad, Dr. Phil. H. Rinks Skrift: Om Grønlænderne M.m.* Copenhagen, 1882.

Hørring, Hugo. *Fortsatte Bemærkninger til Justitsraad Dr. Phils. H. Rinks Skrift: Om Grønlænderne M.m.* Copenhagen, 1883.

Houston, C. Stuart, and John Richardson. *Arctic Ordeal: The Journal of John Richardson, Surgeon-Naturalist with Franklin, 1820–1822.* Montreal: McGill-Queen's Press, 1994.

[Hugh, Murray]. "ART. V.—A Voyage of Discovery, Made under the Orders of the Admiralty, in His Majesty's Ships Isabella and Alexander, for the Purpose of Exploring Baffin's Bay, and Inquiring into the Probability of a North-West Passage." Edited by Francis Jeffrey. *Edinburgh Review, 1802–1929,* March 1819, 337.

Hultén, Eric. *Flora of Alaska and Neighboring Territories: A Manual of the Vascular Plants.* Stanford, CA: Stanford University Press, 1968.

Humboldt, Alexander von. *Des lignes isothermes et de la distribution de la chaleur sur le globe.* Paris: Perronneau, 1817.

Huntington, Patricia A. M. "Robert E. Peary and the Cape York Meteorites." *Polar Geography* 26, no. 1 (January 2002): 53–65.

Hwami, Munyaradzi. "Frantz Fanon and the Problematic of Decolonization: Perspectives on Zimbabwe." *African Identities* 14, no. 1 (January 2016): 19–37.

"Ichyologiske Bidrag til den Grönlandske Fauna." *Det Kongelige Danske Videnskabernes Selskabs Skrifter: Naturvidenskabelig og Mathematisk Afdeling,* 7 (Copenhagen: Bianco Lunos, 1838), 93.

Innis, Harold Adams. *The Fur Trade in Canada: An Introduction to Canadian Economic History.* Rev. ed. Toronto: University of Toronto Press, 1999.

Innis, Harold Adams. "The Importance of Staple Products." In *Approaches to Canadian Economic History: A Selection of Essays,* edited by William Thomas Easterbrook and Mel Watkins, 16–19. Toronto: Carleton University Press, [1984] 2003.

Jackson, Roland. "Eunice Foote, John Tyndall and a Question of Priority." *Notes and Records: The Royal Society Journal of the History of Science,* February 28, 2019. https://doi-org.ezp.lib.cam. ac.uk/10.1098/rsnr.2018.0066.

Jameson, Robert. "Literary and Scientific Intelligence." *Edinburgh Magazine and Literary Miscellany,* 1817, 367–69.

Jensen, J. A. D. "Vandring Paa den Grønlandske Indlandsis i Aaret 1878." *Geografisk Tidsskrift* 3 (1879): 100–107.

Jensen, Nils Aage. *Carl—Polarfarer.* Copenhagen: Lindhardt og Ringhof, 2014.

Johnson, R. E. "Biography—Richardson, Sir John." *Dictionary of Canadian Biography.* Vol. 9, *1861–1870.* Accessed July 22, 2016. http://www.biographi.ca/en/bio.php?id_nbr=4670.

Johnstrup, Johannes Frederick, and Hinrich Rink, eds. *Gieseckes mineralogiske rejse i Grønland.* Copenhagen: B. Lunos bogtrykkeri, 1878.

Johnstrup, Johannes Frederick, and Japatus Steenstrup, eds. "Karl Ludwig Gieseckes Mineralogisches Reisejournal über Grönland 1806–13." *In Meddelelser om Grønland,* 1910, 35:1–478.

Joshi, Priti. "Race." In *Charles Dickens in Context,* edited by Sally Ledger and Holly Furneaux, 292–300. Cambridge: Cambridge University Press, 2011.

Jowitt, Claire, and Carey Daniel, eds. *Richard Hakluyt and Travel Writing in Early Modern Europe.* Farnham, UK: Ashgate, 2012.

Kaalund, Bodil. *The Art of Greenland: Sculpture, Crafts, Painting.* Translated by Kenneth Tindall. Berkeley: University of California Press, 1983.

Kaalund, Nanna Katrine Lüders. "A Frosty Disagreement: John Tyndall, James David Forbes, and the Early Formation of the X-Club." *Annals of Science* 74, no. 4 (October 2017): 282–98.

Kane, Elisha Kent. *Arctic Explorations: The Second Grinnell Expedition in Search of Sir John Franklin, 1853, '54, '55.* 2 vols. Philadelphia: Childs & Peterson, 1857.

Keighren, Innes M., Charles W. J. Withers, and Bill Bell. *Travels into Print: Exploration, Writing, and Publishing with John Murray, 1773–1859.* Chicago: University of Chicago Press, 2015.

Kemp, Martin. *Seen/Unseen: The Visual Ideas behind Art and Science.* Oxford: Oxford University Press, 2006.

King, Richard. "The Arctic Expeditions." *Athenaeum,* December 11, 1847, 1273–74.

King, Richard. *The Franklin Expedition from First to Last.* London: John Churchill, 1855.

Knudsen, Henning. *Fortællingen om Flora Danica.* Copenhagen: Statens Naturhistoriske Museum; Lindhardt og Ringhof, 2014.

Koerbel, Hermann F. "Karl Weyprecht." In *Encyclopedia of the Arctic,* edited by Mark Nuttall, 2172–73. New York: Routledge, 2012.

Koerner, Lisbet. *Linnaeus: Nature and Nation.* Cambridge, MA: Harvard University Press, 2009.

Kohler, Robert E. *Landscapes and Labscapes: Exploring the Lab-Field Border in Biology.* Chicago: University of Chicago Press, 2002.

Kommissionen for Ledelsen af de Geologiske og Geografiske Undersøgelser i Grønland. "Indberetning til Indenrigsministeriet om Undersøgelserne i Aarene 1876, 1877 og 1878." In *Meddelelser om Grønland,* 1879, 1:3–15.

Kommissionen for Ledelsen af de Geologiske og Geografiske Undersøgelser i Grønland. "Indberetning til Indenrigsministeriet om Undersøgelserne i Aarene 1878, 1879 og 1880." In *Meddelelser om Grønland,* 1881, 2:211–220.

Kommissionen for Ledelsen af de Geologiske og Geografiske Undersøgelser i Grønland. *Meddelelser om Grønland.* Copenhagen: C. A. Reitzels Forlag, 1879–.

Kommissionen for Ledelsen af de Geologiske og Geografiske Undersøgelser i Grønland. "Undersøgelserne i Aarene 1878–80 Paa Vestkysten af Grönland, Indberetning til Indenrigsministeriet." *Geografisk Tidsskrift* 5 (January 1881): 58–61.

Konishi, Shino, Maria Nugent, and Tiffany Shellam. *Indigenous Intermediaries: New Perspectives on Exploration Archives.* ANU Press, 2015. www.jstor.org/stable/j.ctt19705zg.

Kornerup, Andreas Nicolaus. "Geologiske Iagttagelser fra Vestkysten af Grønland." In *Meddelelser om Grønland,* 1879, 1:80.

Kraus, Alexander. "Scientists and Heroes: International Arctic Cooperation at the End of the Nineteenth Century." *New Global Studies* 7, no. 2 (2013): 101–16.

Krech, Shepard, III, ed. *The Subarctic Fur Trade: Native Social and Economic Adaptations.* Vancouver: UBC Press, 2011.

Krüger, Tobias. *Discovering the Ice Ages: International Reception and Consequences for a Historical Understanding of Climate.* First English ed. Translated by Ann M. Hentschel. Leiden: Brill, [2008] 2013.

Kühle, Ebbe. *Danmarks Historie i et Globalt Perspektiv.* Copenhagen: Gyldendal, 2008.

Kuklick, Henrika. "Personal Equations: Reflections on the History of Fieldwork, with Special Reference to Sociocultural Anthropology." *Isis* 102, no. 1 (2011): 1–33.

Kuklick, Henrika, and Robert E. Kohler. "Introduction." *Osiris* 11 (1996): 1–14.

Lankford, Kelly. "Arctic Explorer Robert E. Peary's Other Quest: Money, Science, and the Year 1897." *American Nineteenth Century History* 9, no. 1 (March 2008): 37–60.

Lanone, Catherine. "Arctic Romance under a Cloud: Franklin's Second Expedition by Land (1825–7)." In *Arctic Exploration in the Nineteenth Century: Discovering the Northwest Passage*, edited by Frédéric Regard, 95–114. London: Pickering and Chatto; Pittsburgh: University of Pittsburgh Press, 2015.

Latta, Jeffrey Blair. *The Franklin Conspiracy: An Astonishing Solution to the Lost Arctic Expedition.* Toronto: Dundurn Press, 2001.

Laudan, Rachel. *From Mineralogy to Geology: The Foundations of a Science, 1650–1830.* Chicago: University of Chicago Press, 1987.

Laughton, John Knox. "An Address Delivered at the Annual General Meeting, January 16th, 1884." *Quarterly Journal of the Royal Meteorological Society* 10, no. 50 (April 1884): 77–87.

Launer, Donald. *Navigation through the Ages.* New York: Sheridan House, 2009.

LeMoine, Genevieve. "Elatu's Funeral: A Glimpse of Inughuit-American Relations on Robert E. Peary's 1898–1902 Expedition." *Arctic* 67, no. 3 (2014): 340–46.

Lentin, Alana. *Racism and Ethnic Discrimination.* New York: Rosen Publishing Group, 2011.

Levere, Trevor H. *Science and the Canadian Arctic: A Century of Exploration, 1818–1918.* Cambridge: Cambridge University Press, 2004.

Lidegaard, Mads. "Hans Hendrik Fra Fiskenæsset." *Grønland* 8 (August 1968): 249–56.

Lightman, Bernard. *Victorian Popularizers of Science: Designing Nature for New Audiences.* Chicago: University of Chicago Press, 2009.

Lindgren, Raymond E. *Norway-Sweden: Union, Disunion, and Scandinavian Integration.* Princeton, NJ: Princeton University Press, 1959.

Lindsay, Debra J. *Science in the Subarctic: Trappers, Traders, and the Smithsonian Institution.* Washington, DC: Smithsonian Institution Press, 1993.

Livingstone, David N. *Putting Science in Its Place: Geographies of Scientific Knowledge.* Chicago: University of Chicago Press, 2010.

Lloyd, Christopher. *Mr. Barrow of the Admiralty: A Life of Sir John Barrow.* London: Irvington Publishers, 1970.

"Log Book." *Geographical Magazine* (April 1876): 104–5.

Løve, Jan. *Hans Hendrik og Hans Ø: Beretningen om Hans Hendrik og de to Hans Øer.* Copenhagen: Det Grønlandske Selskab, 2016.

MacLaren, I. S. "The Aesthetic Map of the North, 1845–1859." *Arctic* 38, no. 2 (June 1985): 89–103.

Mann, Michael. "'Torchbearers upon the Path of Progress': Britain's Ideology of a 'Moral and Material Progress' in India; An Introductory Essay." In Fischer-Tiné and Mann, *Colonialism as Civilizing Mission*, 1–26.

Manning, Patrick, and Jerry H. Bentley. "The Problem of Interactions in World History." *American Historical Review* 101, no. 3 (1996): 771.

Marchildon, Gregory P. *The Early Northwest.* Regina, SK: University of Regina Press, 2008.

[Markham, Clements Robert]. "The Arctic Campaign of 1873." *Ocean Highways: The Geographical Record* 1, no. 3 (1874): 89–91.

Markham, Clements Robert. *The Threshold of the Unknown Region.* London: Sampson Low, Marston, Low, and Searle, 1873.

Marquardt, Ole. "Between Science and Politics: The Eskimology of Hinrich Johannes Rink." In *Early Inuit Studies: Themes and Transitions, 1850s–1980s*, edited by Igor Krupnik, 35–54. Washington, DC: Smithsonian Institution Press, 2016.

Marquardt, Ole. "Greenland's Demography, 1700–2000: The Interplay of Economic Activities and Religion." *Études/Inuit/Studies* 26, no. 2 (2002): 47–69.

Martin-Nielsen, Janet. *Eismitte in the Scientific Imagination: Knowledge and Politics at the Center of Greenland*. New York: Palgrave Macmillan, 2013.

Mazlish, Bruce. "Comparing Global History to World History." *Journal of Interdisciplinary History* 28, no. 3 (1998): 385–95.

McClellan, James Edward, III, and Harold Dorn. *Science and Technology in World History: An Introduction*. Rev. ed. Baltimore: Johns Hopkins University Press, [1999] 2006.

McClintock, Anne. *Imperial Leather: Race, Gender, and Sexuality in the Colonial Contest*. New York: Routledge, 1995.

McClintock, Francis Leopold. "Discoveries by the Late Expedition in Search of Sir John Franklin and His Party." *Proceedings of the Royal Geographical Society of London* 30 (1860): 2–14.

M[c]'Clintock, Francis Leopold. *The Voyage of the "Fox" in the Arctic Seas: A Narrative of the Discovery of the Fate of Sir John Franklin and His Companions*. London: John Murray, 1859.

McCorristine, Shane, and Victoria Herrmann. "The 'Old Arctics': Notices of Franklin Search Expedition Veterans in the British Press, 1876–1934." *Polar Record* 52, no. 2 (March 2016): 215–29.

McGoogan, Kenneth. *Fatal Passage: The Story of John Rae, the Artic Hero Time Forgot*. New York: Carroll & Graf, 2002.

Melville, George. *In the Lena Delta: A Narrative of the Search for Lieut-Commander De Long and His Companions, Followed by an Account of the Greely Relief Expedition and a Proposed Method of Reaching the North Pole*. Boston: Houghton Mifflin, 1884.

Merton, Thomas (pseud). "Arctic Natural History." *Literary Magnet of the Belles Lettres, Science, and the Fine Arts, 1824–1826* 1, no. 1 (January 1824): 51–54.

Moore, Grace. *Dickens and Empire: Discourses of Class, Race and Colonialism in the Works of Charles Dickens*. Aldershot, UK: Ashgate, 2004.

Morin, Karen M. *Civic Discipline: Geography in America, 1860–1890*. New York: Routledge, 2016.

Morrow, Marina, Olena Hankivsky, and Colleen Varcoe, eds. *Women's Health in Canada: Critical Perspectives on Theory and Policy*. Toronto: University of Toronto Press, 2008.

Murchison, Roderick Impey. *Address to the Royal Geographical Society of London; Delivered at the Anniversary Meeting, May 25th, 1857*. London: W. Clowes and Sons, 1857.

Murphy, David. *The Arctic Fox: Francis Leopold-McClintock*. Toronto: Dundurn, 2004.

Murray-Miller, Gavin. *The Cult of the Modern: Trans-Mediterranean France and the Construction of French Modernity*. Lincoln: University of Nebraska Press, 2017.

Nansen, Fridtjof. *Paa Ski over Grønland: En Skildring af den Norske Grønlands-Ekspedition 1888–89*. H. Aschehoug, 1890.

Nares, George Strong. *Narrative of a Voyage to the Polar Sea: During 1875–6 in H. M. Ships 'Alert' and 'Discovery.'* 2 vols. London: Slow, Marston, Searle, & Rivington, 1878.

Nares, George Strong. *The Official Report of the Recent Arctic Expedition*. London: John Murray, 1876.

"Narrative of an Expedition to the Shores of the Arctic Sea in 1846 and 1847." *Athenaeum*, July 27, 1850, 784–85.

"Narrative of an Expedition to the Shores of the Arctic Sea, in 1846 and 1847." *Quarterly Review*, March 1853, 386–421.

"Narrative of a Second Voyage in Search of a North-West Passage, and of a Residence in the Arctic Regions, during the Years 1829, 1830, 1831, 1832, 1833, by Sir John Ross, C.B., K.S.A., K.C.S., &c. &c." *Literary Gazette: A Weekly Journal of Literature, Science, and the Fine Arts*, May 9, 1835, 289–92.

"Narrative of the Discoveries on the North Coast of America, Effected by the Officers of the Hudson's Bay Company, during the Years 1836–9." *Critic of Literature, Art, Science, and the Drama, 1843–1844*, February 1844, 85–86.

Neill, Deborah. *Networks in Tropical Medicine: Internationalism, Colonialism, and the Rise of a Medical Specialty, 1890–1930*. Stanford, CA: Stanford University Press, 2012.

Neill, Robin. *A History of Canadian Economic Thought*. Routledge History of Economic Thought Series. London: Routledge, 1991.

Nellemann, George. "Hinrich Rink and Applied Anthropology in Greenland in the 1860's." *Human Organization* 28, no. 2 (June 1969): 166–74.

Neumayer, Georg. "Die geographische Probleme innerhalb der Polarzonen in ihrem inneren zusammenhange Beleuchtet [Intrinsic Aspects of Geographical Problems within Polar Regions]." *Hydrographische Mittheilungen* 2, no. 5–7 (1874): 51–53.

Nielsen, Niels Kayser. "MYTE: Sagde Dalgas 'Hvad udad Tabes, Skal indad Vindes'?" Danmarks Historien, Aarhus University. Accessed September 17, 2016. http://danmarkshistorien.dk/leksikon-og-kilder/vis/materiale/myte-sagde-dalgas-hvad-udad-tabes-skal-indad-vindes/?no_cache=1.

O'Connor, Ralph. *The Earth on Show: Fossils and the Poetics of Popular Science, 1802–1856*. Chicago: University of Chicago Press, 2008.

O'Dochartaigh, Eavan. "The Visual Culture of the Franklin Search Expeditions to the Arctic (1848–55)." PhD dissertation, National University of Ireland, Galway, 2018.

Oldendow, Knud. *Grønlændervennen Hinrich Rink: Videnskabsmand, Skribent og Grønlandsadministrator*. Det Grønlandske Selskabs Skrifter 18. Copenhagen: Det Grønlandske Selskab, 1955.

Oldroyd, David Roger. *Thinking about the Earth: A History of Ideas in Geology*. Cambridge, MA: Harvard University Press, 1996.

Osterhammel, Jürgen, and Niels P. Petersson. *Globalization: A Short History*. Translated by Dona Geyer. Princeton, NJ: Princeton University Press, 2005.

"Our Weekly Gossip." *Athenaeum*, September 5, 1840, 701–2.

Panneton, Daniel, and Leslie H. Neatby. "John Rae." *The Canadian Encyclopedia*. Accessed December 19, 2016. http://www.thecanadianencyclopedia.com/en/article/john-rae/.

Petersen, Carl. *Den Sidste Franklin-Expedition med "Fox," Capt. M'Clintock, ved Carl Petersen*. Copenhagen: Fr. Woldikes Forlagsboghandel, 1860.

Pickering, Andrew. *Constructing Quarks: A Sociological History of Particle Physics*. Chicago: University of Chicago Press, 1999.

Pingel, C. "XXIX. W.A. Graah, Undersøgelsesreise til Østkysten af Grønland." *Maanedsskrift for Litteratur* 10 (1833): 593–648.

"Polar Expedition." *Imperial Magazine*, August 1819, 697–703.

Porter, Theodore M. *Trust in Numbers: The Pursuit of Objectivity in Science and Public Life*. Princeton, NJ: Princeton University Press, 1995.

"Postnyheder," *Kalundborg Avis*, October 1859, 1.

Potter, Russell A. *Arctic Spectacles: The Frozen North in Visual Culture, 1818–1875*. Seattle: University of Washington Press, 2007.

Potter, Russell A. *Finding Franklin: The Untold Story of a 165-Year Search*. Montreal: McGill-Queen's Press, 2016.

Potter, Russell A. "Introduction: Exploration and Sacrifice; The Cultural Logic of Arctic Discovery." In *Arctic Exploration in the Nineteenth Century: Discovering the Northwest Passage*, edited by Frédéric Regard, 1–18. London: Pickering and Chatto; Pittsburgh: University of Pittsburgh Press, 2015.

Potter, Russell A., and Douglas W. Wamsley. "The Sublime Yet Awful Grandeur: The Arctic Panoramas of Elisha Kent Kane." *Polar Record* 35, no. 194 (July 1999): 193–206.

Powell, Richard C. "Becoming a Geographical Scientist: Oral Histories of Arctic Fieldwork." *Transactions of the Institute of British Geographers*, n.s., 33, no. 4 (October 2008): 548–65.

Pratt, Mary Louise. *Imperial Eyes: Travel Writing and Transculturation*. London: Routledge, 1992.

Priebe, Janina. "From Siam to Greenland: Danish Economic Imperialism at the Turn of the Twentieth Century." *Journal of World History* 27, no. 4 (2016): 619–40.

Qureshi, Sadiah. *Peoples on Parade: Exhibitions, Empire, and Anthropology in Nineteenth-Century Britain*. Chicago: University of Chicago Press, 2011.

R. S. "ART. VI.—The Journal of the Royal Geographical Society of London." *London and Westminster Review, Apr. 1836–Mar. 1840*, August 1838, 273–392.

Rae, John. *Narrative of an Expedition to the Shores of the Arctic Sea, in 1846 and 1847*. London: T. & W. Boone, 1850.

Ratcliff, Jessica. *The Transit of Venus Enterprise in Victorian Britain*. London: Pickering and Chatto; Pittsburgh: University of Pittsburgh Press, 2008.

Ray, Arthur J., and Donald B. Freeman. *"Give Us Good Measure": An Economic Analysis of Relations between the Indians and the Hudson's Bay Company before 1763*. Toronto: University of Toronto Press, 1978.

Reidy, Michael S. *Tides of History: Ocean Science and Her Majesty's Navy*. Chicago: University of Chicago Press, 2009.

Reidy, Michael S., Gary R. Kroll, and Erik M. Conway. *Exploration and Science: Social Impact and Interaction*. Santa Barbara, CA: ABC-CLIO, 2007.

Reinhardt, J[ohannes]. "Ichyologiske Bidrag til den Grönlandske Fauna." *Det Kongelige Danske Videnskabernes Selskabs Skrifter: Naturvidenskabelig og Mathematisk Afdeling*, no. 7 (Copenhagen: Bianco Lunos, 1838): 83–196.

Rich, Edwin Ernest. *Hudson's Bay Company 1670–1870*. Vol. 1, *1821–1870*. 3 vols. New York: Macmillian, 1961.

Richardson, John. *Arctic Searching Expedition: A Journal of a Boat-Voyage through Rupert's Land and the Arctic Sea, in Search of the Discovery Ships under Command of Sir John Franklin. With an Appendix on the Physical Geography of North America*. 2 vols. London: Longman, Brown, Green and Longmans, 1851.

Richardson, John, ed. *Fauna Boreali-Americana, or, The Zoology of the Northern Parts of British America: Containing Descriptions of the Objects of Natural History Collected on the Late Northern Land Expeditions, under Command of Captain Sir John Franklin, R.N.* 4 vols. London: John Murray, 1829–37.

Rink, Hinrich Rink. *Danish Greenland, Its People and Its Products*. Edited by Robert Brown. London: H. S. King, 1877.

Rink, Hinrich. "Introduction." In Hendrik, *Memoirs of Hans Hendrik*, 1–20.

Rink, Hinrich. *Naturhistoriske Bidrag til en Beskrivelse af Grønland*. Copenhagen: L. Kleins Bogtrykkeri, 1857.

Rink, Hinrich. "Nogle Bemærkninger om de Nuværende Grønlænderes Tilstand." *Geografisk Tidsskrift* 1 (January 1877).

Rink, Hinrich. *Om Grønlænderne, deres Fremtid, og de til deres Bedste Sigtende Foranstaltninger*. Copenhagen: Høst, 1882.

Rink, Hinrich. "On the Supposed Discovery, by Dr. E. K. Kane, U. S. N., of the North Coast of Greenland, and of an Open Polar Sea, &c.; As Described in 'Arctic Explorations in the Years

1853, 1854, 1855.'" Translated by Dr. Shaw. *Journal of the Royal Geographical Society of London* 28 (1858): 272–87.

Rink, Hinrich. "On the Supposed Discovery of the North Coast of Greenland and an Open Polar Sea; The Great 'Humboldt Glacier,' and Other Matters Relating to the Formation of Ice in Greenland, as Described in 'Arctic Explorations in the Years 1853–4–5', by Elisha Kent Kane." *Proceedings of the Royal Geographical Society of London* 2, no. 4 (1858): 195–201.

Rink, Hinrich. "The Recent Danish Explorations in Greenland and Their Significance as to Arctic Science in General." *Proceedings of the American Philosophical Society* 22, no. 120 (1885): 280–96.

Rink, Hinrich. "Udsigt over Nordgrönlands Geognosi, Især med Hensyn til Bjergmassernes Mineralogiske Sammensætning." In *Om den Geographiske Beskaffenhed af de Danske Handelsdistriker i Nordgrönland, Tilligemed en Udsigt over Nordgrönlands Geognosi*, 35–62. Copenhagen: B. Lunos Kgl. Hof-Bogtrykkeri, 1852.

Ross, John. *Explanation and Answer to Mr. John Braithwaite's Supplement to Captain Sir John Ross's Narrative of a Second Voyage in the Victory, in Search of a Northwest Passage*. London: A. W. Webster, 1835.

Ross, John. *An Explanation of Captain Sabine's Remarks on the Late Voyage of Discovery to Baffin's Bay*. London: John Murray, 1819.

Ross, John. *A Treatise on Navigation by Steam: Comprising a History of the Steam Engine, and an Essay towards a System of the Naval Tactics Peculiar to Steam Navigation, as Applicable Both to Commerce and Maritime Warfare; Including a Comparison of Its Advantages as Related to Other Systems in the Circumstances of Speed, Safety and Economy, but More Particularly in That of the National Defence*. London: Longman, Rees, Orme, Brown, and Green, 1828.

Ross, John. *A Voyage of Discovery, Made under the Orders of the Admiralty, in His Majesty's Ships Isabella and Alexander, for the Purpose of Exploring Baffin's Bay, and Inquiring into the Probability of a North-West Passage*. London: John Murray, 1819.

Ross, John, and James Clark Ross. *Narrative of a Second Voyage in Search of a North-West Passage, and of a Residence in the Arctic Regions during the Years 1829, 1830, 1831, 1832, 1833*. London: A. W. Webster, 1835.

Ross, Maurice James. *Polar Pioneers: John Ross and James Clark Ross*. Montreal: McGill-Queen's Press, 1994.

Ross, W. Gillies. "The Admiralty and the Franklin Search." *Polar Record* 40, no. 4 (October 2004): 289–301.

Ross, W. Gillies. "False Leads in the Franklin Search." *Polar Record* 39, no. 2 (April 2003): 131–60.

Ross, W. Gillies. "The Type and Number of Expeditions in the Franklin Search 1847–1859." *Arctic* 55, no. 1 (2002): 57–69.

"Ross's Expedition." *Chambers's Edinburgh Journal, Feb. 1832–Dec. 1853*, November 28, 1835, 346–47.

Rostgaard, Marianne, and Lotte Schou. *Kulturmøder i Dansk Kolonihistorie*. Copenhagen: Gyldendal Uddannelse, 2010.

Rothenberg, Marc. "Making Science Global? Coordinated Enterprises in Nineteenth-Century Science." In *Globalizing Polar Science: Reconsidering the International Polar and Geophysical Years*, edited by Roger D. Launius, James Rodger Fleming, and David H. DeVorkin, 23–35. Palgrave Studies in the History of Science and Technology. New York: Palgrave Macmillan, 2010.

Royal Geographical Society of Great Britain, ed. *Arctic Geography and Ethnology: A Selection of Papers on Arctic Geography and Ethnology. Reprinted, and Presented to the Arctic Expedition of 1875, by the President, Council, and Fellows of the Royal Geographical Society*. London: John Murray, 1875.

Royal Geographical Society of Great Britain. "The President's Address on Presenting Medals." *Journal of the Royal Geographical Society* 9 (1839): ix–xii.

Royal Geographical Society of Great Britain, ed. "Sessions 1872–73." *Proceedings of the Royal Geographical Society of London*, 1873, 77.

"The Royal Society." *Nature* 27 (December 1882): 162–67.

Rudwick, Martin J. S. *The Great Devonian Controversy: The Shaping of Scientific Knowledge among Gentlemanly Specialists.* Chicago: University of Chicago Press, 1988.

Rudwick, Martin J. S. *The Meaning of Fossils: Episodes in the History of Palaeontology.* Chicago: University of Chicago Press, 1976.

Ruggles, Richard. *A Country So Interesting: The Hudson's Bay Company and Two Centuries of Mapping, 1670–1870.* Montreal: McGill-Queen's Press, 1991.

Sabine, Edward. *Remarks on the Account of the Late Voyage of Discovery to Baffin's Bay.* London: R. and A. Taylor, 1819.

Sandler, Martin W. *Resolute: The Epic Search for the Northwest Passage and John Franklin, and the Discovery of the Queen's Ghost Ship.* New York: Sterling, 2008.

Schröder, W. "The First International Polar Year (1882–1883) and International Geophysical Cooperation." *Earth Sciences History* 10, no. 2 (January 1991): 223–26.

Schwartz, Stuart B., ed. *Implicit Understandings: Observing, Reporting and Reflecting on the Encounters between Europeans and Other Peoples in the Early Modern Era.* Cambridge: Cambridge University Press, 1994.

"Science Notices." *Dublin Review, 1836–1910; London*, April 1883, 463–67.

Scoresby, William. *The Arctic Whaling Journals of William Scoresby the Younger: The Voyages of 1817, 1818 and 1820.* Edited by C. Ian Jackson. 3 vols. New York: Routledge, 2009.

Scoresby, William. *Journal of a Voyage to the Northern Whale-Fishery: Including Researches and Discoveries on the Eastern Coast of West Greenland.* Edinburgh: Archibald Constable, 1823.

Secord, James A. "Knowledge in Transit." *Isis; an International Review Devoted to the History of Science and Its Cultural Influences* 95, no. 4 (December 2004): 654–72.

Secord, James A. *Victorian Sensation: The Extraordinary Publication, Reception, and Secret Authorship of Vestiges of the Natural History of Creation.* Chicago: University of Chicago Press, 2000.

Selskabet for Danmarks Kirkehistorie. *Kirkehistoriske Samlinger.* Copenhagen: Akademisk Forlag, 1911.

Sera-Shriar, Efram. "Arctic Observers: Richard King, Monogenism and the Historicisation of Inuit through Travel Narratives." *Studies in History and Philosophy of Science Part C: Studies in History and Philosophy of Biological and Biomedical Sciences* 51 (June 2015): 23–31.

Sera-Shriar, Efram, ed. *Historicizing Humans: Deep Time, Evolution, and Race in Nineteenth-Century British Sciences.* Pittsburgh: University of Pittsburgh Press, 2018.

Sera-Shriar, Efram. *The Making of British Anthropology, 1813–1871.* London: Pickering and Chatto; Pittsburgh: University of Pittsburgh Press, 2013.

Sera-Shriar, Efram. "What Is Armchair Anthropology? Observational Practices in 19th-Century British Human Sciences." *History of the Human Sciences* 27, no. 2 (2014): 26–40.

Sevaldsen, Jørgen. "'No Proper Taste for the English Way of Life': Danish Perceptions of Britain 1870–1940." In *Britain and Denmark: Political, Economic and Cultural Relations in the 19th and 20th Centuries*, edited by Jørgen Sevaldsen, 61–72. Aarhus, Denmark: Museum Tusculanum Press, 2003.

Shapin, Steven. *The Scientific Life: A Moral History of a Late Modern Vocation.* Chicago: University of Chicago Press, 2009.

Shapin, Steven, and Simon Schaffer. *Leviathan and the Air-Pump: Hobbes, Boyle, and the Experimental Life.* Rev. ed. Princeton, NJ: Princeton University Press, [1985] 2011.

Sheets-Pyenson, Susan. "'Pearls before Swine': Sir William Dawson's Bakerian Lecture of 1870." *Notes and Records of the Royal Society of London* 45, no. 2 (1991): 177–91.

Shellam, Tiffany. "Miago and the 'Great Northern Men': Indigenous Histories from In-Between." In *Indigenous Mobilities: Across and Beyond the Antipodes*, edited by Rachel Standfield, 185–208. ANU Press, 2018. www.jstor.org/stable/j.ctv3o1dn7.12.

Simpson, Alexander. *The Life and Travels of Thomas Simpson: The Arctic Discoverer.* London: R. Bentley, 1845.

Simpson, Thomas. *Narrative of the Discoveries on the North Coast of America: Effected by the Officers of the Hudson's Bay Company during the Years 1836–39.* London: R. Bentley, 1843.

"Sir John Richardson's Arctic Expedition." *Dublin University Magazine*, April 1852, 458–76.

"Sir John Richardson's Arctic Searching Expedition." *Spectator*, November 15, 1851, 1096–97.

Sivasundaram, Sujit. "Natural History Spiritualized: Civilizing Islanders, Cultivating Breadfruit, and Collecting Souls." *History of Science* 39, no. 4 (December 2001): 417–43.

Sivasundaram, Sujit. *Nature and the Godly Empire: Science and Evangelical Mission in the Pacific, 1795–1850.* Cambridge: Cambridge University Press, 2005.

Sivasundaram, Sujit. "Sciences and the Global: On Methods, Questions, and Theory." *Isis* 100, no. 1 (March 2010): 146–58.

Sivasundaram, Sujit. "Towards a Critical History of Connection: The Port of Colombo, the Geographical 'Circuit,' and the Visual Politics of New Imperialism, ca. 1880–1914." *Comparative Studies in Society and History* 59, no. 2 (April 2017): 346–84.

Steenstrup, K. J. V. "Dr. Phil. Hinrich Johannes Rink." *Geografisk Tidsskrift* 12 (January 1894): 162–66.

Stepan, Nancy Leys. *Picturing Tropical Nature.* Ithaca, NY: Cornell University Press, 2001.

Stephens, George. Translator's note. In Hendrik, *Memoirs of Hans Hendrik.* 20.

Stern, Pamela R. *Daily Life of the Inuit.* Santa Barbara, CA: Greenwood, 2010.

Stewart, D. A. "Sir John Richardson Surgeon, Physician, Sailor, Explorer, Naturalist, Scholar." *British Medical Journal* 1, no. 3654 (1931): 110–12.

Stocking, George W. *The Ethnographer's Magic and Other Essays in the History of Anthropology.* Madison: University of Wisconsin Press, 1992.

Stocking, George W. *Observers Observed: Essays on Ethnographic Fieldwork.* Madison: University of Wisconsin Press, 1984.

Summerhayes, Colin P. "International Collaboration in Antarctica: The International Polar Years, the International Geophysical Year, and the Scientific Committee on Antarctic Research." *Polar Record* 44, no. 4 (October 2008): 321–34.

"Supplement to Captain Sir John Ross's Narrative of a Second Voyage in the Victory, in Search of a North-West Passage, Containing the Suppressed Facts Necessary to a Proper Understanding of the Causes of the Failure of the Steam Machinery of the Victory, and a Just Appreciation of Captain Sir John Ross's Character as an Officer and a Man." *Literary Gazette: A Weekly Journal of Literature, Science, and the Fine Arts*, November 7, 1835, 712.

Swartz, Marvin. *Politics of British Foreign Policy in the Era of Disraeli and Gladstone.* New York: St. Martin's Press, 1985.

Tammiksaar, Erki, Natalia G. Sukhova, and Ian R. Stone. "Russia and the International Polar Year, 1882–1883." *Polar Record* 45, no. 3 (July 2009): 215–23.

"Thomas Simpson, Esq." *Gentleman's Magazine: And Historical Review, July 1856–May 1868*, November 1840, 548–49.

Thomson, Don W. *Men and Meridians: The History of Surveying and Mapping in Canada*. Vol. 3, *1966–69*. Ottawa: R. Duhamel, Queen's printer, 1969.

Thorsen, Svend, and Tage Kaarsted. *De Danske Ministerier: Et Hundred Politisk-Historiske Biografier. [Udg. af Pensionsforsikringsanstalten i Anledning af dens 50 ärs Jubilaeum]*. Nyt Nordisk Forlag, 1967.

Thulesius, Olav. *The Man Who Made the Monitor: A Biography of John Ericsson, Naval Engineer*. Jefferson, NC: McFarland, 2007.

Thyvold, Hans Olav. *Fridtjof Nansen: Explorer, Scientist and Diplomat*. Translated by James Anderson. N.p.: Font Forlag, 2012.

Toft, Peter A., and Inge Høst Seiding. "Circumventing Colonial Policies: Consumption and Family Life as Social Practices in the Early Nineteenth-Century Disko Bay." In *Scandinavian Colonialism and the Rise of Modernity: Small Time Agents in a Global Arena*, edited by Magdalena Naum and Jonas M. Nordin, 105–30. New York: Springer Science & Business Media, 2013.

Topham, Jonathan R. "Beyond the 'Common Context': The Production and Reading of the Bridgewater Treatises." *Isis* 89, no. 2 (June 1998): 233–62.

Topham, Jonathan R. "Scientific Publishing and the Reading of Science in Nineteenth-Century Britain: A Historiographical Survey and Guide to Sources." *Studies in History and Philosophy of Science Part A* 31, no. 4 (2000): 559–612.

Truth and Reconciliation Commission of Canada. *Canada's Residential Schools: The Métis Experience; The Final Report of the Truth and Reconciliation Commission of Canada*. McGill-Queen's Native and Northern Series 83. Montreal: McGill-Queen's University Press, 2016.

Twells, A. *The Civilising Mission and the English Middle Class, 1792–1850: The "Heathen" at Home and Overseas*. Dordrecht, Netherlands: Springer, 2008.

"Udenlandske Efterretninger." *Den til Forsendelse med de Kongelige Brevposter Privilegerede Berlingske Politiske og Avertissementstidende*, October 28, 1854, 2.

"Udlandet." *Fyens Stiftstidende*, October 30, 1854, 1.

Vallgårda, Karen. *Imperial Childhoods and Christian Mission: Education and Emotions in South India and Denmark*. Basingstoke, UK: Palgrave Macmillan, 2014.

Wallace, Hugh N. *The Navy, the Company, and Richard King: British Exploration in the Canadian Arctic, 1829–1860*. Montreal: McGill-Queen's University Press, 1980.

Wamsley, Douglas W. "'We Are Fully in the Expedition': Philadelphia's Support for the North Greenland Expeditions of Robert E. Peary, 1891–1895." *Geographical Review* 107, no. 1 (January 2017): 207–35.

Watson, Annette. "William Scoresby." In *Encyclopedia of the Arctic*, edited by Mark Nuttall, 1850–51. New York: Routledge, 2012.

Weld, Charles Richard. *Arctic Expeditions*. London: John Murray, 1850.

Wells, John Campion. *The Gateway to the Polynia: A Voyage to Spitzbergen*. London: Henry S. King, 1873.

Weyprecht, Karl. "Fundamental Principles of Arctic Investigation." Paper presented at the meeting of the Association of the German Naturalists and Physicians, Graz, September 18, 1875.

Weyprecht, Karl. "Fundamental Principles of Scientific Arctic Investigation." Paper presented at the annual meeting of the Academy of Science, Vienna, January 18, 1875.

Whittaker, A. "The Travels and Travails of Sir Charles Lewis Giesecke." In *Four Centuries of Geological Travel: The Search for Knowledge on Foot, Bicycle, Sledge and Camel*, edited by Patrick Wyse Jackson, 149–60. London: Geological Society of London, 2007.

Wilder, Gary. "Race, Reason, Impasse: Cesaire, Fanon, and the Legacy of Emancipation." *Radical History Review* 90, no. 1 (August 2004): 31–61.

Williamson, Oliver E. *The Economic Institutions of Capitalism*. New York: Free Press, 1985.

Wilson, Michael D. *Writing Home: Indigenous Narratives of Resistance*. East Lansing: Michigan State University Press, 2008.

Withers, Charles W. J., and Innes M. Keighren. "Travels into Print: Authoring, Editing and Narratives of Travel and Exploration, c.1815—c.1857." *Transactions of the Institute of British Geographers*, n.s., 36, no. 4 (October 2011): 560–73.

Woitkowitz, John. "Science, Networks, and Knowledge Communities: August Petermann and the Construction of the Open Polar Sea." Paper presented at the annual meeting of the Canadian Historical Association, Vancouver, June 5, 2019.

Woitkowitz, John, and Nanna Kaalund. "August Petermann, Elisha Kent Kane, and the Making of the Open Polar Sea." Paper presented at the Science Museum Seminar Series, London, June 18, 2019.

Woodman, David C. *Unravelling the Franklin Mystery: Inuit Testimony*. 2nd ed. Montreal: McGill-Queen's Press, 2015.

Wyss, Hilary E. *Writing Indians: Literacy, Christianity, and Native Community in Early America*. Pbk. ed. Amherst: University of Massachusetts Press, [2000] 2003.

Yandell, Keith E. *Faith and Narrative*. Oxford: Oxford University Press, 2001.

Zeller, Suzanne. "Humboldt and the Habitability of Canada's Great Northwest." *Geographical Review* 96, no. 3 (2006): 382–98.

Zhou, Yong. *The Histories of the International Polar Years and the Inception and Development of the International Geophysical Year*. 1st ed. Annals of the International Geophysical Year, vol. 1. London: Pergamon, 1959.

INDEX